HOW TO READ SHOP DRAWINGS

WITH SPECIAL REFERENCE TO ARC WELDING

Welding Symbols as Standardized
by The American Welding Society

Adapted by Mark Prosser, M.A.Ed., CWI/CWE
Welding/Fabrication Instructor,
Blackhawk Technical College, Janesville, Wisconsin.

Published by
The James F. Lincoln Arc Welding Foundation.
P.O. Box 17188, Cleveland, Ohio 44117-1199

THE JAMES F. LINCOLN ARC WELDING FOUNDATION

Printed in Taunton, Massachusetts

SIMPLE BLUEPRINT READING
FIRST EDITION, July 1940
Reprinted, November 1940
Reprinted, March 1941

SECOND EDITION, October 1941
Reprinted, April 1942
Reprinted, August 1942
Reprinted, March 1943
Reprinted, January 1947

THIRD EDITION, June 1948
Reprinted, January 1950
Reprinted, March 1951

FOURTH EDITION, January 1953
Reprinted, June 1956
Reprinted, March 1958

HOW TO READ SHOP DRAWINGS
FIRST EDITION, January 1961
Reprinted, January 1962
Reprinted, August 1963
Reprinted, January 1965
Reprinted, July 1966
Reprinted, November 1967
Revised, November 1972
Reprinted, October 1973
Revised, May 1979
Reprinted, September 1981
Reprinted, October 1986
SECOND EDITION, June 2016

LIBRARY OF CONGRESS CONTROL NUMBER: 61013942
ISBN Number: 978-0-937390-08-5

CONTENTS

Introduction.. 1-4

Chapter 1: Functions of Blueprints 5-16

Chapter 2: Understanding Welding Symbols 17-36

Chapter 3: Measuring and Tools 37-46

Chapter 4: Welding Processes 47-58

Chapter 5: Fillet Welds 59-70

Chapter 6: Groove Welds 71-86

Chapter 7: Plug, Slot and Edge Welds 87-98

Chapter 8: Spot, Seam, Projection & Surfacing Welds 99-112

Chapter 9: Industry Materials 113-126

Chapter 10: Weld Inspection & Examination 127-144

Glossary 145-151

Conclusion.................................. 152

Introduction

Blueprints come in many different configurations, and different symbol systems have been developed over the years to convey information accurately and precisely throughout the entire manufacturing process. A machinist's blueprint is very different from a welding blueprint. The symbols used for welding prints may not exist on a machinist's print. Regardless of type, prints serve the same purpose for each trade: to relay information in a quick and accurate way that everyone understands, from concept, to engineering and design, to the welder building the product on the shop floor.

In the welding industry, there is often a disconnect between the designers and engineers and the welders. Welding blueprints use a specific symbol system, which needs to be learned by everyone involved in making the product. When welders read welding blueprints, they expect to see symbols showing weld type, size, location, length, process, joint type and often procedural steps used to produce a high quality product; therefore, it is essential that the designer or engineer understand the weld symbol system so they can convey accurate and needed information to the welder.

Good blueprints can be one of the most cost effective aspects of any manufacturing facility. If prints are created with the welder in mind and provide the specific information the welder needs to create a product, it can lead to greater production, less rework, faster turnaround and even better worker morale. When a welder spends less time trying to figure out a blueprint, they spend more time producing. Taking the time up front to create excellent prints will pay off exponentially when the print makes it to the welder's hands.

The Goal of the Book

The goal of this book is not to baffle you with all of the different types of prints used in the trades. To sufficiently cover each type of print would require a much larger book and would involve a great deal of information that a welder may never use. In this text we focus on the welding trade and the symbols, information, and notations used on welding blueprints. This book begins with basic information such as types of lines and continues with more complex information such as multistep welds with different joint configurations. This book contains all the vital information for the welding trade. The information is presented in sequence, each chapter building on the previous chapter, to ensure the reader or student understands one concept before moving to the next.

This book is for high school students, college students, apprentice programs, or anyone else who needs the ability to understand welding blueprints. It would also be beneficial to designers and engineers who need to understand how to relay their information to welders. This book presents the information in a simple, easily understood manner that can benefit even the most competent print reader.

This book has been developed in accordance with the most recent AWS (American Welding Society) and ANSI (American National Standards Institute) information and standards. Each chapter contains review questions and exercises to reinforce the information presented in the chapter. Completing each exercise will help the reader to solidify the information and develop a deeper understanding. There are also real prints in the book that reflect what readers would see in a "real life" production environment. Some of the prints have been developed exclusively for this book, and others are actual shop prints from various manufacturing plants.

Most of the information in the book can apply to any welding situation, however, prints can be more specific because each company has different requirements and standards. The author and publisher of this book have presented the most accurate and recent information and provide no guarantee nor accept any responsibility in any way for how this information might be used by the reader. We expect you will find this book a quick read, easy to understand, and to the point. It will get you on your way to adding "competent print reader" to your list of welding skills.

About the Author

Mark A. Prosser is a welding/fabrication instructor at a Wisconsin technical college. He has worked in many different aspects of the welding industry, ranging from automotive, to structural steel and the piping industry over the last 25 years and has taught for the last 13 years. He has authored five books and has written many welding-related articles for different industry magazines and publications. Mark, together with his partner, nationally renowned custom car and bike builder Bryan Fuller, is the co-owner and author of Full-Bore Publishing, which has produced two books: "Full-Bore Sheet Metal" and "Full-Bore Welding." Mark is also the owner of "Question Mark," a welding educational consulting company. He has been a certified welding inspector (CWI/CWE) and instructor for many years. He earned his Associate's degree in welding technology from Ferris State University, his Bachelor's degree in industrial education from Central Michigan University and his Master's degree in adult education from Jones International University. Mark travels the country performing welding and sheet metal seminars to promote the extremely valuable human trait of working with your hands.

About the Foundation

The James F. Lincoln Foundation was founded in 1936 and has dedicated over 70 years to help people in the welding and fabrication industry. The foundation held its first technical paper contest in 1938 in which nearly 2000 papers were submitted. The foundation awarded over 400 papers cash prizes totaling $200,000.00, equivalent to 2.6 million dollars in today's standards. Lincoln Electric's "Procedure Handbook of Arc Welding," considered by many as the "Bible" of the welding industry, was published in 1933 and has sold over 500,000 copies.

By the 1980's, the foundation became the world's largest publisher of welding literature and continues to make strides to provide updated information to schools, industry and individuals who work in the welding industry worldwide. This book is one of several new initiatives of the foundation to address the lack of print reading skills in the industry.

As the industry advances in technology, lean manufacturing and overall growth, Lincoln Electric and the James F. Lincoln Foundation continue to lead the way into the future of welding on a global level. The dedication to providing information to the industry has been second to none for over 70 years. As we move forward, the James F. Lincoln Foundation will continue to produce educational books, papers and information to help everyone on any level along on their welding path.

Safety Practices in Welding

Important Note:
So that you can protect yourself against hazards, every welder should be familiar with American National Standard ANSI Z49.1, "Safety in Welding and Cutting," and should follow the safety practices in that document. Z49.1 is now available for download at no charge at:

www.lincolnelectric.com/safety

or at the AWS website

http://www.aws.org.

Download and read it!

E205: Arc Welding Safety
A safety document summarizing many of the hazards and safe practices for welding. Download and print your own copy at:

www.lincolnelectric.com/safety

Free copies available from Lincoln Electric at:

https://marcomcentral.app.pti.com/dukeprinting/lincolnusmarketing/home.aspx?uigroup_id=18186

CHAPTER 1: Functions of Blueprints

Chapter Objective

After learning the information in this chapter you will be able to:

- Understand the basic components that make up blueprints
- Identify the functions of prints and drawings
- Identify the different types of lines and their functions
- Compare different views on drawings
- Practice sketching techniques

Key Terms

Object lines
Hidden lines
Center lines
Dimension lines
Extension lines
Leader lines
Cutting plane lines
Sectioning lines
Phantom lines
Multiview drawings
Orthographic drawings
Auxiliary views
Sectional views
Isometric views
Sketching techniques

When I was a child, my mother was extremely efficient in shorthand. For those of you who don't know, shorthand is a series of symbols that represent words.

My mother worked as an office professional long before hand-held recorders, smartphones or personal computers were available. Sometimes I would go to work with her, and I still remember watching her use shorthand to write down everything that everyone said during meetings. At the end of the meeting, her paper would be covered with weird lines and obscure symbols. The amazing thing was that my mother could then read the whole thing back to me; entire conversations were contained in just two or three lines. Welding symbols are the shorthand of the drawing world. Many drawings designed by engineers need to be translated into a language welders can understand. A misunderstanding between the engineer and the welder can drastically affect the workflow. Learning blueprint symbols and understanding what they mean will make you a better welder or fabricator and allow you to understand a wide variety of prints.

Different trades use different blueprints. For example, a machinist's print is different from a welding blue-print. It will contain information that most welders won't understand. Prints designed for welders need to be constructed by someone who understands welding symbols, weld types, weld sizes, weld maps, weld profiles, which process to use and other important information that we welders need in order to produce the desired parts. Engineers who understand blueprints in general but are not necessarily well versed in the language of welding prints have a hard time relaying information to the welder.

Companies must be efficient to be successful in today's competitive manufacturing market. Good prints are crucial to any competitive and profitable welding and fabrication operation. The cost of making the part almost always increases when there is any confusion about the prints. In a fabrication shop, confusion between the design engineer and the welder can lead to a great deal of cost due to overwelding, distortion, rework and welder confusion. Everyone from the design table to the shop floor needs to speak the same language and understand the prints.

Lines

Lines are a great place to start learning how prints work. Believe it or not, there are several different types of lines and each have different, specific meanings. One simple line could show a different surface, the center of a circle or the dimension of the part. Some lines are dashed, some are thick, and some have an arrow or a dot at the end, but each one represents details that help us understand the information being presented. Table 1.1 shows all of the different types of lines and what each line represents.

LINES AND THEIR FUNCTIONS			
	Name of Line	Function	Characteristics
1.	Object Lines	Used to outline the overall shape of a part	Thick: 
2.	Hidden Lines	Used to show hidden features of a part	Thin: 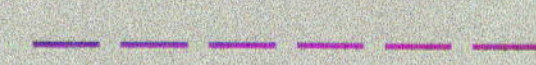
3.	Center Lines	Used to show the center of holes, arcs and circles	Thin: 
4.	Dimension Lines	Used to point to the extension lines and contain measurements in the middle	Thin: 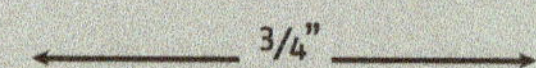
5.	Extension Lines	Used to show the exact point of measurement	Thin: 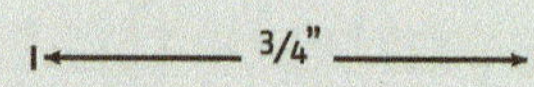
6.	Leader Lines	Used to identify specific features of a part	Thin: 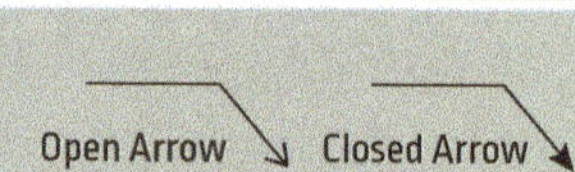
7.	Cutting Plane Lines	Used to show the internal features of a part	Thick: 
8.	Section Lines	Used to identify material types and to show different surfaces	Thin: 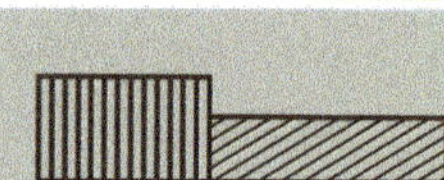
9.	Short Break Lines	Used to show short breaks in a part (freehand)	Thick: 
10.	Long Break Lines	Used to show long breaks in a part (freehand)	Thin: 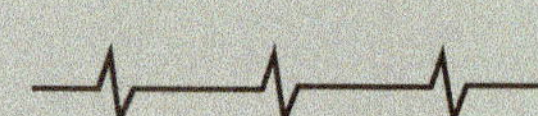
11.	Phantom Lines	Used to show movement of a part	Thin: 

Table 1.1 Each line has a specific function. Learning how they relate to and interact with each other is important.

Object Lines

Object lines are thick, heavy lines that represent the main lines of the outside of the part being constructed. Object lines are used to help us visualize the outline of the part.

Hidden Object Lines

Hidden lines are medium thickness, dashed lines that show the edges of a part that are not visible to the eye. Hidden lines give you the X-ray vision you need to see edges of the part while, at the same time, looking through the part.

Centerlines

Centerlines are thin lines composed of alternating long and short dashes. Centerlines are used to show the center of holes, arcs and circles.

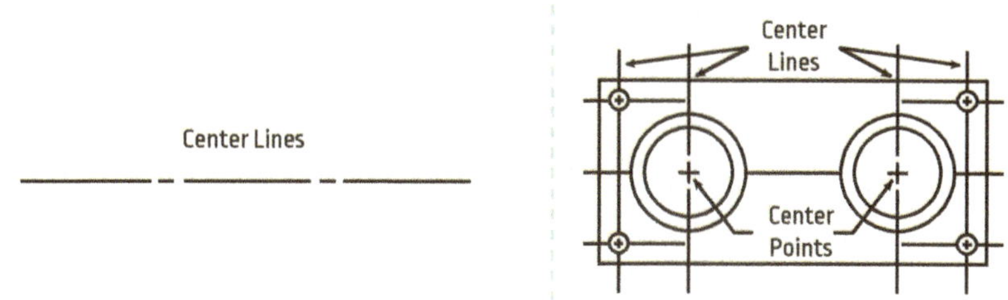

Dimension Lines

Dimension lines are thin, solid lines with arrowheads on each end. They are used in conjunction with extension lines (see below) to show the measurements of the part. Dimension lines touch the extension lines and have the relevant measurement written in the middle of the line.

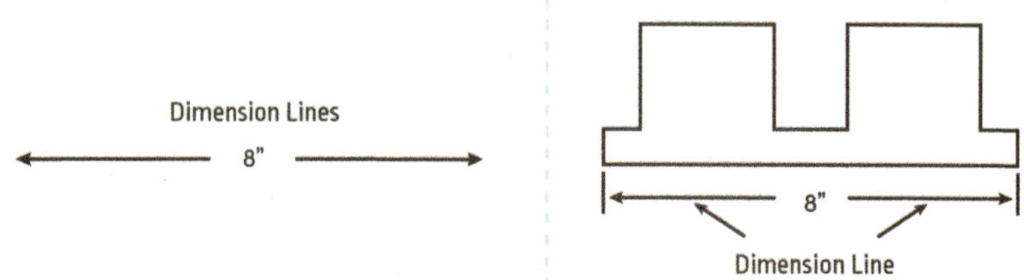

Extension Lines

Extension lines are thin lines used with dimension lines, as discussed above. One extension line appears at either end of the dimension line to show where the dimension lines' arrowheads end.

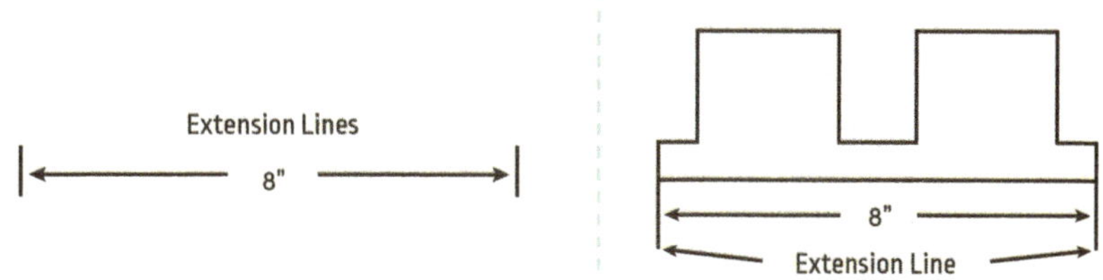

Leader Lines

A leader line is a thin, straight line with an arrow on one end. Some leader lines have a solid dot on the end instead of an arrow. A leader line points to the surface of a part to show a specific detail, dimension or a special note.

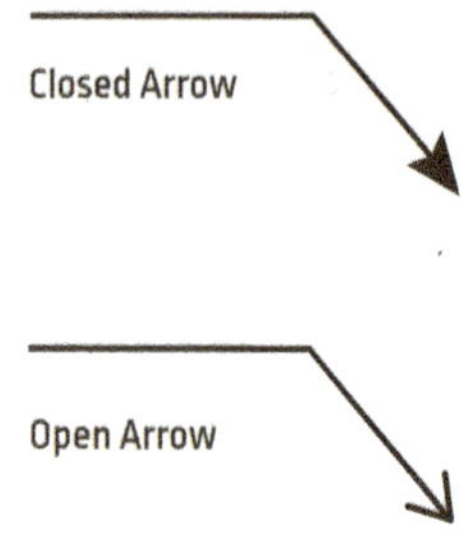

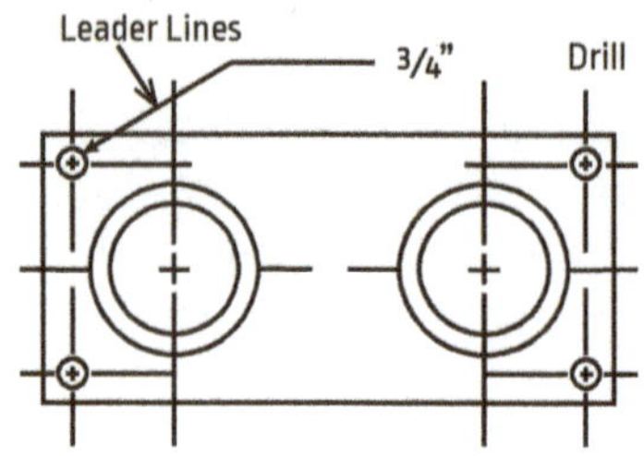

Cutting Plane Lines

Cutting plane lines are heavy, dashed lines made up of sets of one long dash and two short dashes. These lines end in 90° angles topped with arrows. They show where a cut through the part will be made. The arrows indicate the direction from which we should look at the part.

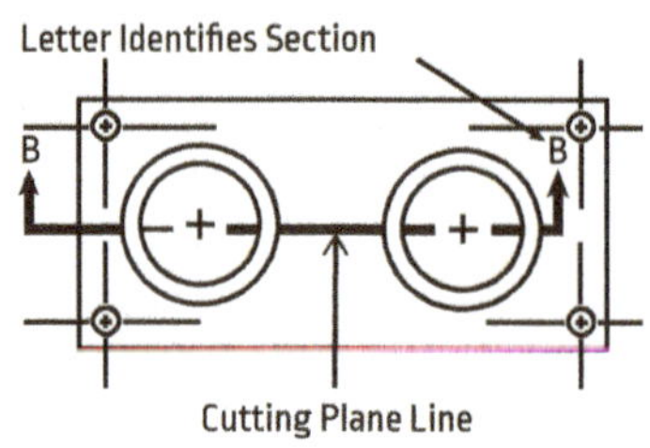

Section Lines

Section lines are lines configured in several ways to show different materials such as steel, aluminum, copper, cast iron and any other material you may use. Section lines are used with cutting plane lines to show different types of materials when the material specifications are shown somewhere else. Section lines are also known as crosshatching.

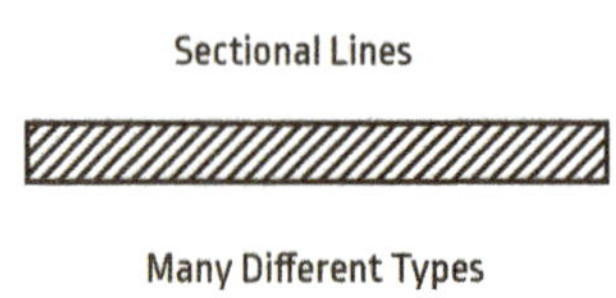

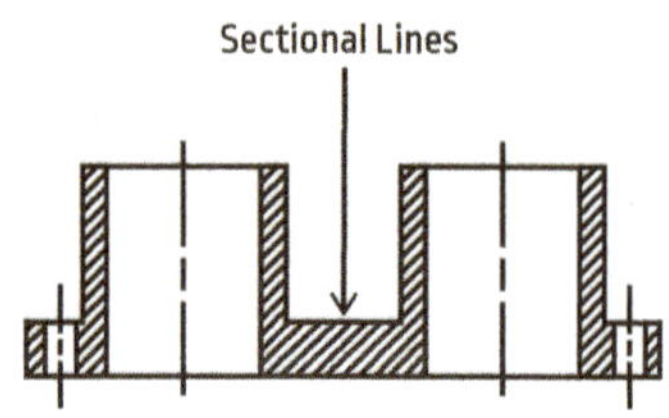

Break Lines

Break lines are generally used to shorten long parts so they can fit on the page. Short break lines are heavier and long break lines are thin. They both look like a zigzag line that cuts the part in half. When break lines are used on round tube, the zigzag becomes more of an S shape, exposing the inside of the pipe or tube.

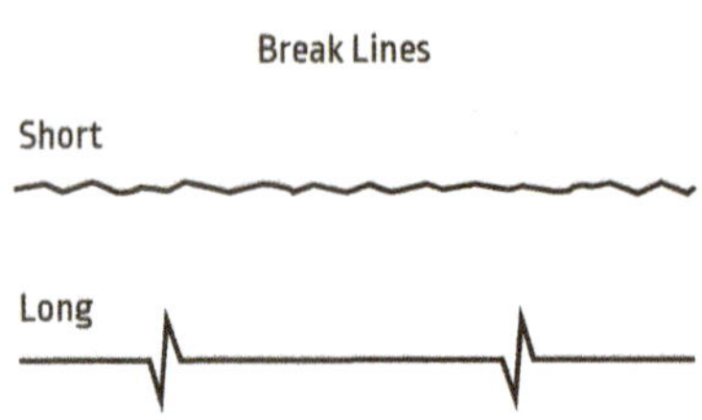

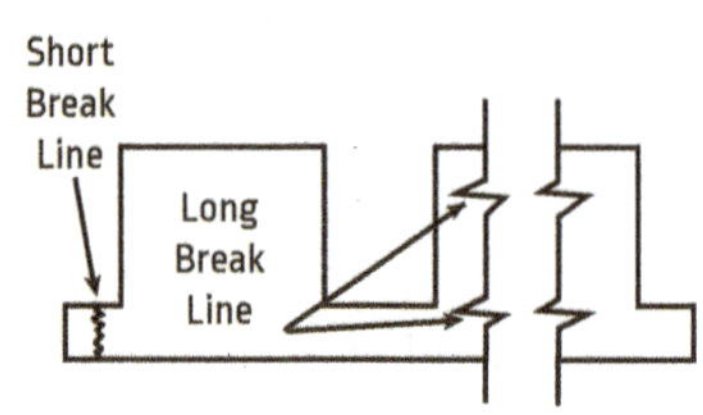

Phantom Lines

Phantom lines show the different positions of a part that moves. Phantom lines are light, dashed lines made up of one long and two short dashes.Phantom lines also show details like threaded parts or rounded corners. They are always related to the motion of a part that moves in and out repeatedly, like a hydraulic ram.

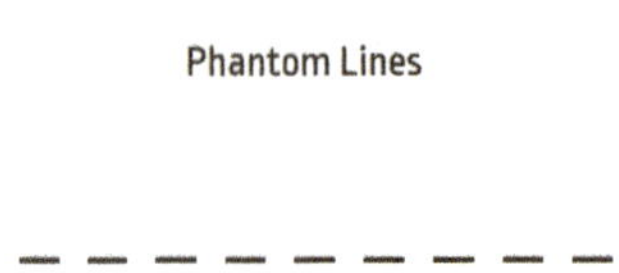

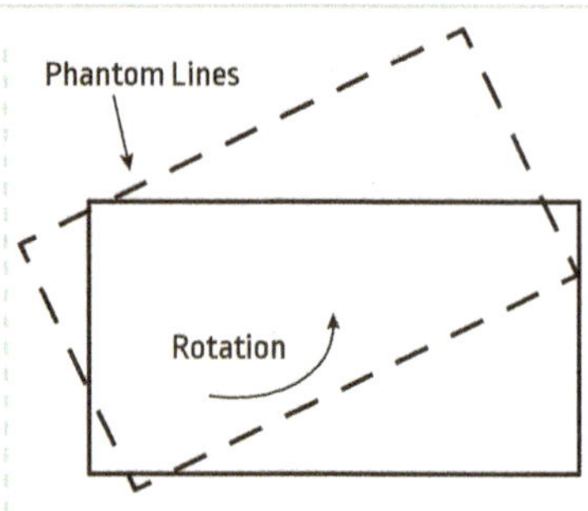

Multiview Drawings

Multiview drawings are composed of more than one view of a part. They show us each side of a part and help us imagine the construction process of the part. Multiview drawings usually contain three views of a part: front view, right side view and top view (Fig. 1.1). In almost all cases, these three views will provide all the information we need to fully visualize the part. The front view of the part usually gives the best representation of the part's shape and major details.

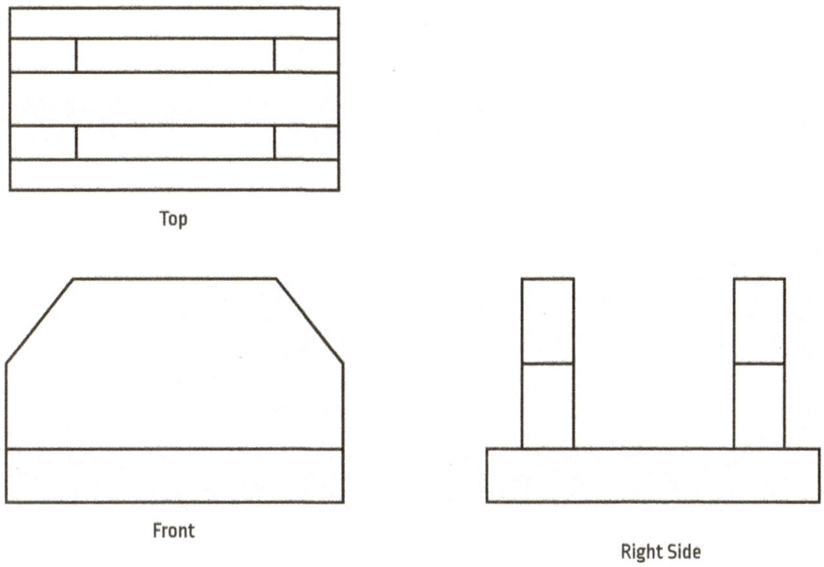

Figure 1.1 Examples of 3 views: top, front, right side of a simple part.

These three views will usually provide all the information we need to build the part, however, sometimes we need additional views to illustrate the part more clearly and make dimensioning easier. These additional views may contain a bottom view, left side view or a rear view (Fig. 1.2).

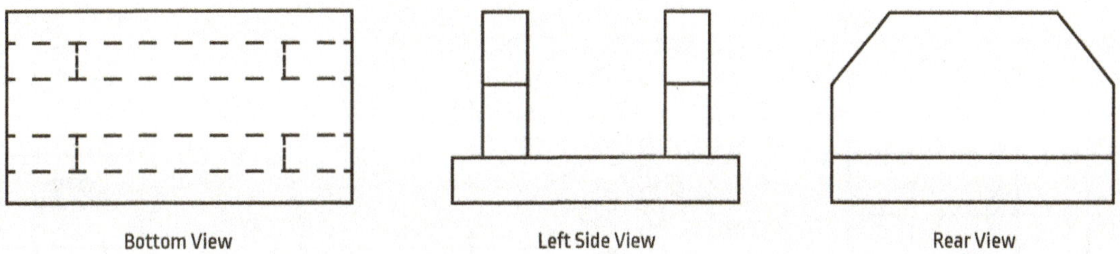

Figure 1.2 Example of all views of the same part.

Orthographic Projection

An orthographic projection helps us understand the relationship between a physical part and the drawing of the part. Understanding this relationship takes time and practice. To visualize orthographic projection, imagine a part in a clear box. We can examine the part from six different views, one side at a time. If we look straight at each side of the box, we can see the front, top and right side view, and we can begin to understand the relationship between those three views and a 3-D view of the part. If we imagine that all sides of the box have hinges that allow us to unfold the box and lay it flat on our paper, we can begin to see how the orthographic projection corresponds to different views on a drawing.

Auxiliary Views

The front, side and top views don't always provide a true representation of the size of an object or part with slanted or angular sides. Many times a slanted side of a part will require an additional, or auxiliary, view.

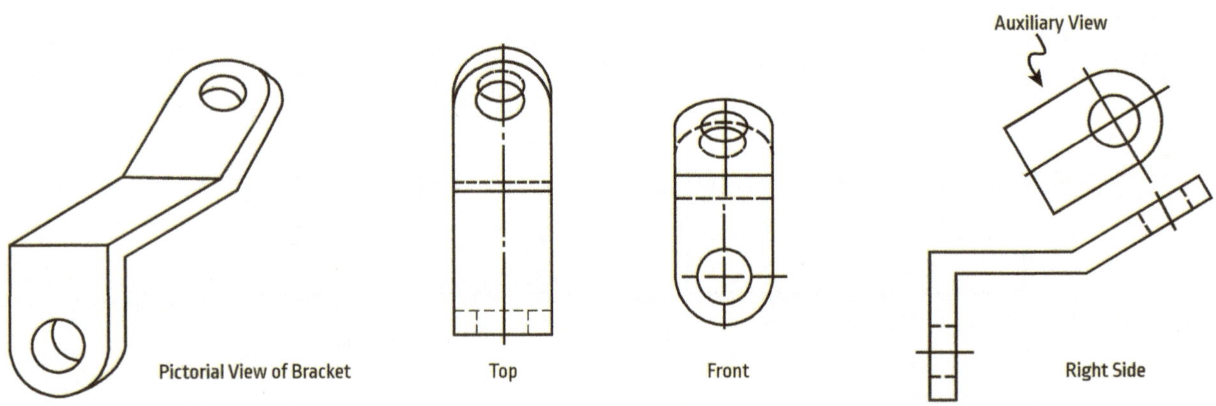

Figure 1.3 A part with a slanted side that can't be seen from three views.

An auxiliary view will show different sloped surfaces of the part that may not be seen in the front, top or side view. Auxiliary views also show a difference in width of the slanted surface (Fig. 1.4).

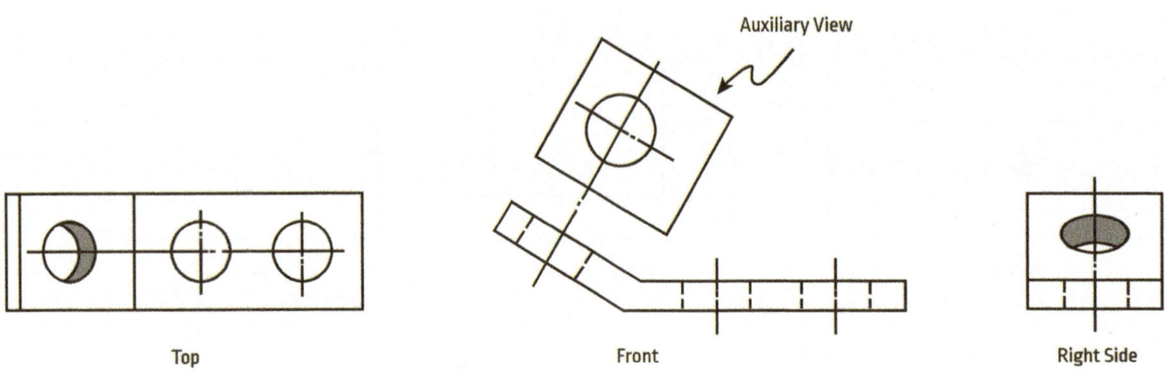

Figure 1.4 A part with a slanted side that changes width, which can't be seen from three views.

Additional auxiliary views are used to show the true size and shape of the angular surface. These views are always projected at a 90° angle from the primary or front view (Fig. 1.5).

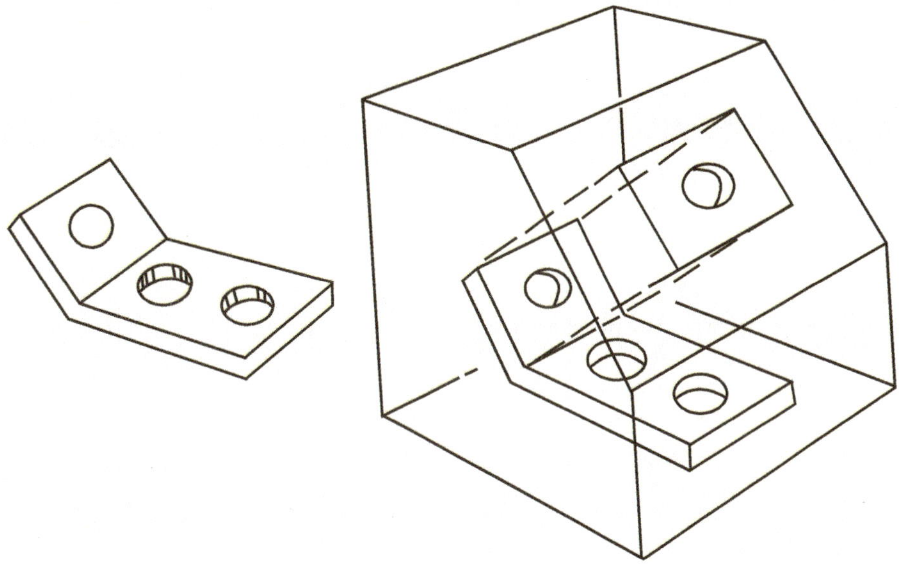

Figure 1.5 Transferring the view from the front view 90° to show slanted surface.

Sectional Views

Sectional views let us take a look inside more complex parts. Without sectional views, we would need multiple hidden lines to show the insides of a complex part, which could lead to a confusing drawing. A sectional view shows us what the part would look like inside if we were to use a cutting plane line to cut the part in half perpendicular to our line of sight of the part (Fig. 1.6).

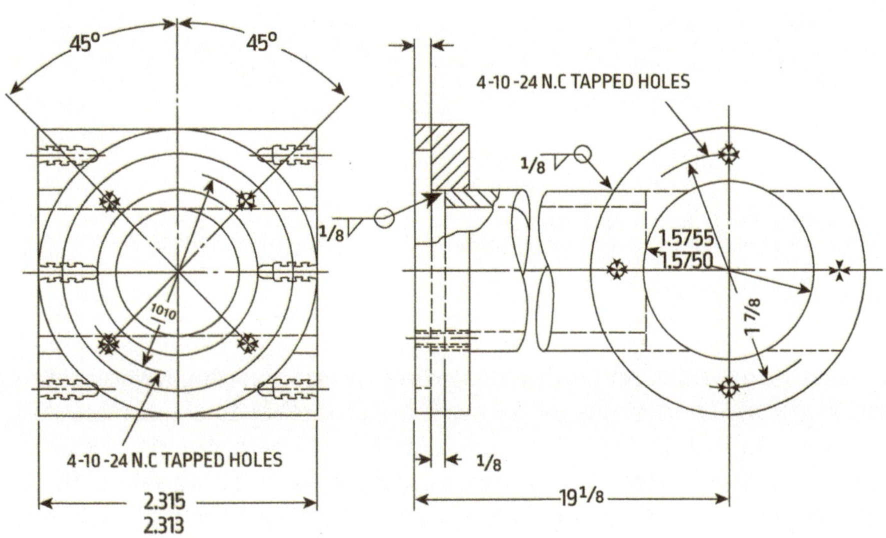

Figure 1.6 Part being cut with cutting plane line.

Isometric Views

An isometric view is a three-dimensional view directly related to the front, side and top view of a drawing. The isometric view shows the part as a whole, configured so that we can see two sides and the top at the same time. The isometric view shows the complete part in a functional way that makes drawing the three views easier.

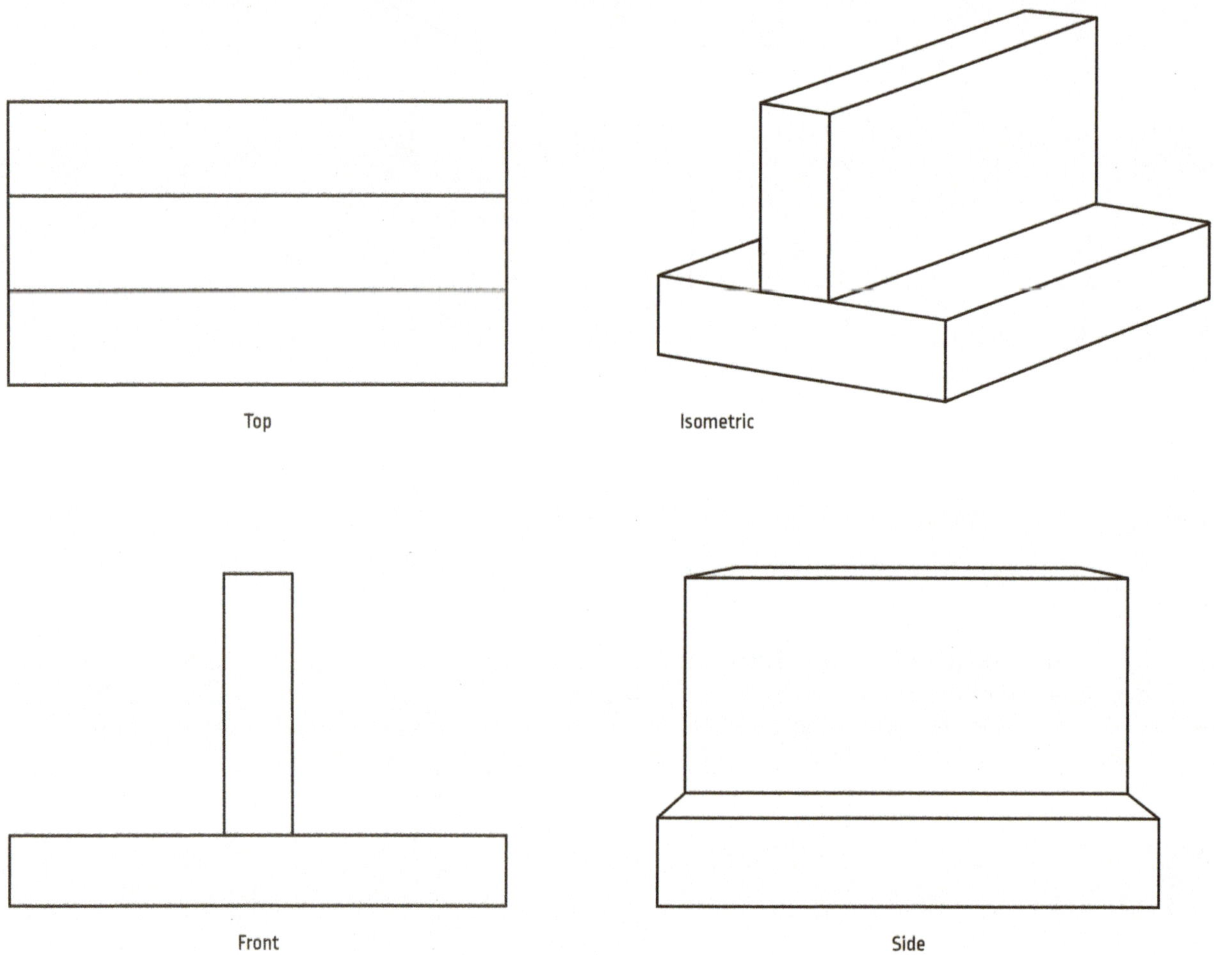

Figure 1.7 Three views with isometric view of the part for reference.

Reading Drawings

Many welders learn to read prints through getting help from others, reading a good reference book and studying all of the different components of a good drawing. Remember to ask for help when needed and always pay attention to the smallest details. Try to visualize the shape of the parts to be assembled by breaking the information into small pieces and understanding what is being asked. In your free time, practice drawing and sketching any parts you see. Not only will this improve your sketching techniques, but it will also develop your ability to envision the part more easily. Learning to read prints can be frustrating at first, but with a little practice and effort, reading prints will become second nature for you.

Sketching

A sketch is a simple freehand drawing that is used to quickly depict a part or an assembly. It is the fastest and easiest way to put ideas onto paper. I once had the chance to watch Chip Foose from the TV show "Overhauling" hand sketch a three-dimensional drawing of a '56 Chevrolet. The quickness with which he put his ideas onto paper was nothing short of amazing. The drawing was then refined with tighter lines, more accurate dimensions and better versions until it was a fully drawn artistic rendering.

Many drawings begin with a simple sketch that is enhanced and refined to make it more accurate and descriptive. Using hand sketching techniques on the shop floor is an excellent way for you to capture your ideas. Dimensions and details can be quickly added to sketches to flesh out your vision. Sketching will take a good deal of practice, but it is a valuable skill that will allow you to easily relate ideas with nothing more than a pencil and a piece of paper.

Be sure to start your sketches with light lines, not dark, heavy lines. This will allow you to pick the best line and darken it. It will also make it easy to erase less desirable lines. The following exercises will help you see the relationship between the different types of views and lines and help you hone your sketching techniques.

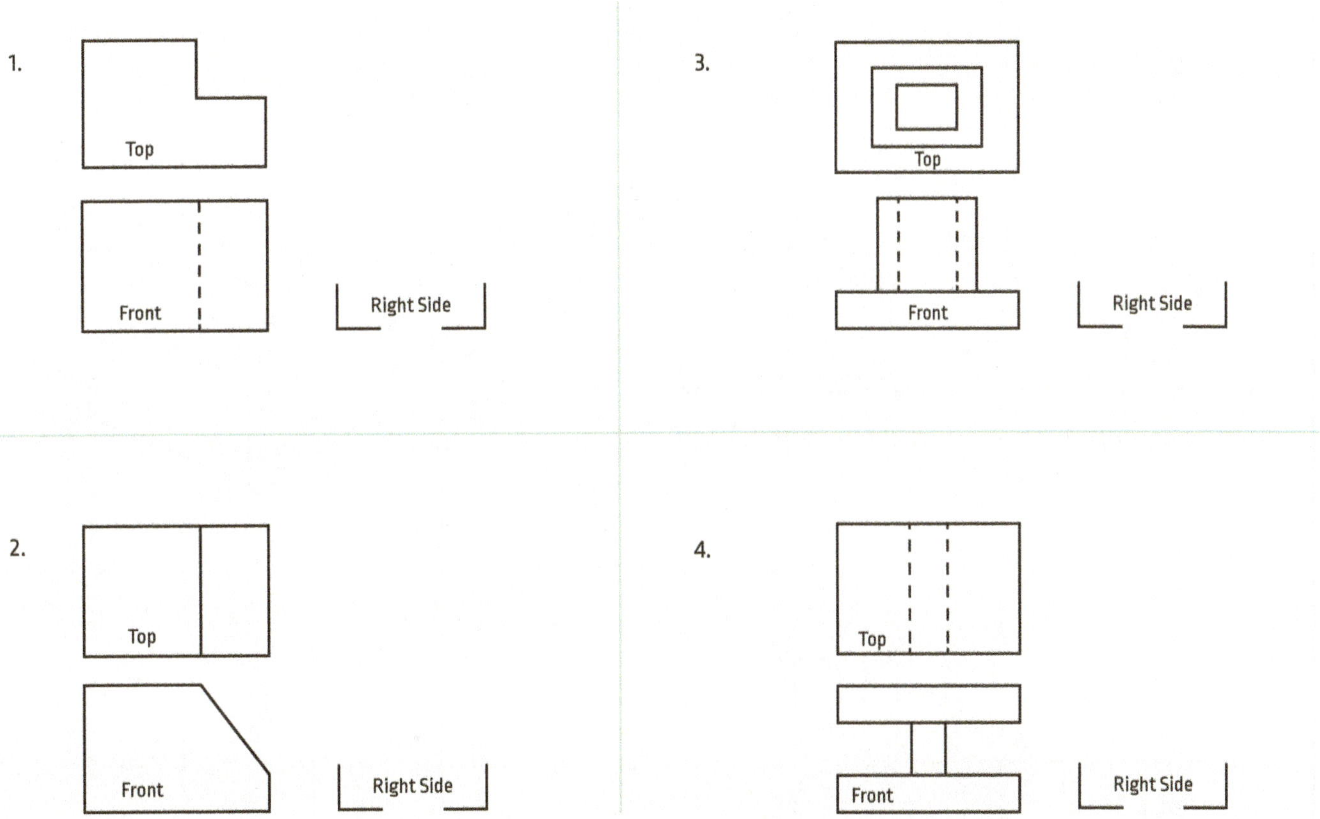

Figure 1.8 The top and front views provide enough information to sketch the right side view. Sketch the right side view using the other views.

Review Exercises - Measurement and Line Identification

Directions: Use the drawing below to identify the different measurements and different line types.

1. ____________________
2. ____________________
3. ____________________
4. ____________________
5. ____________________
6. ____________________
7. ____________________
8. ____________________
9. ____________________
10. ____________________
11. ____________________
12. ____________________
13. ____________________
14. ____________________
15. ____________________
16. ____________________

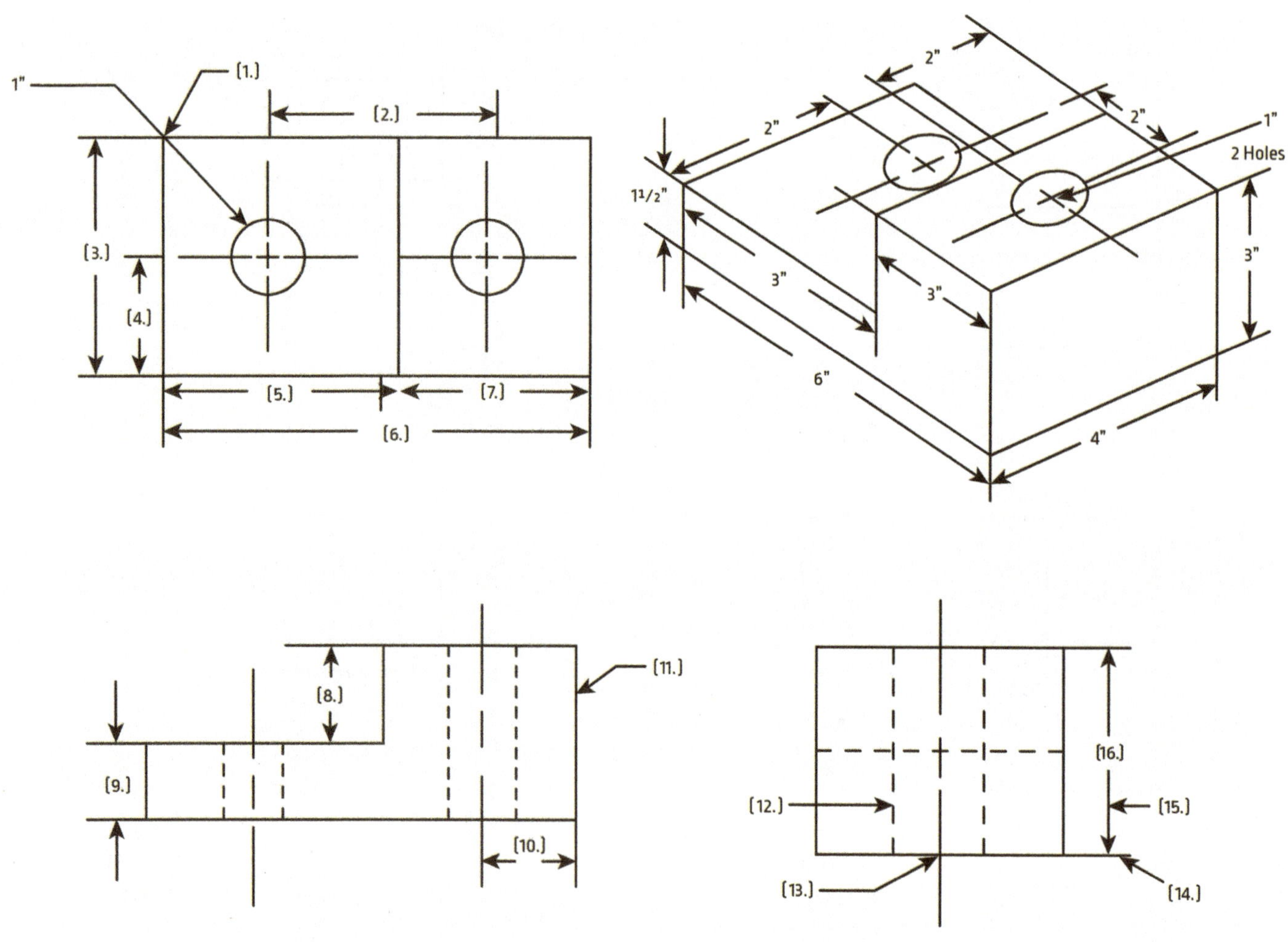

Review Exercises - Sketching

Directions: Sketch the required views using the given drawings.

1.

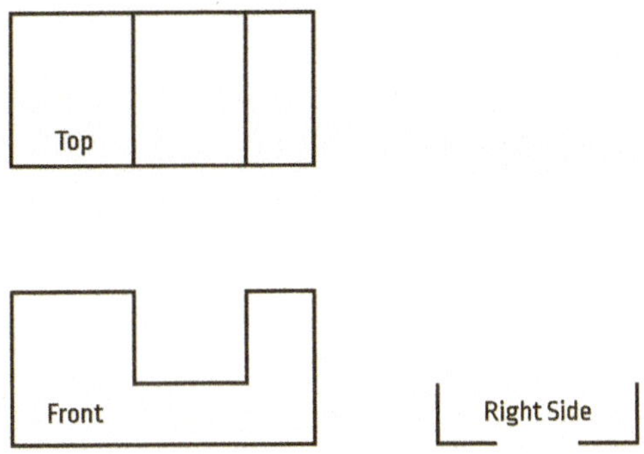

2.

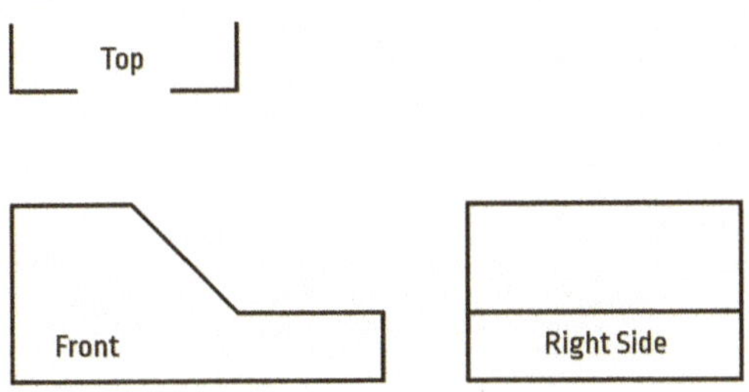

3.

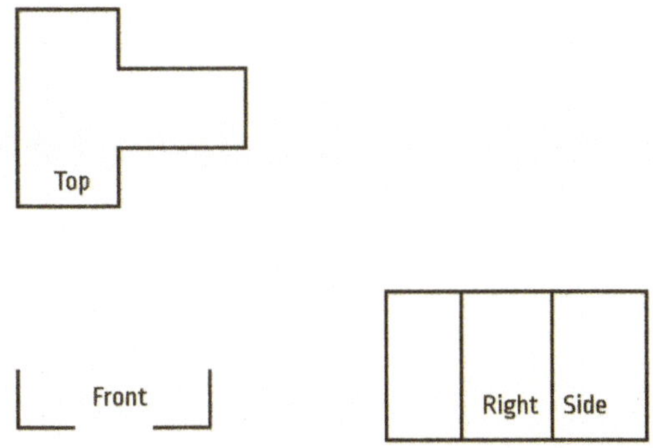

4.

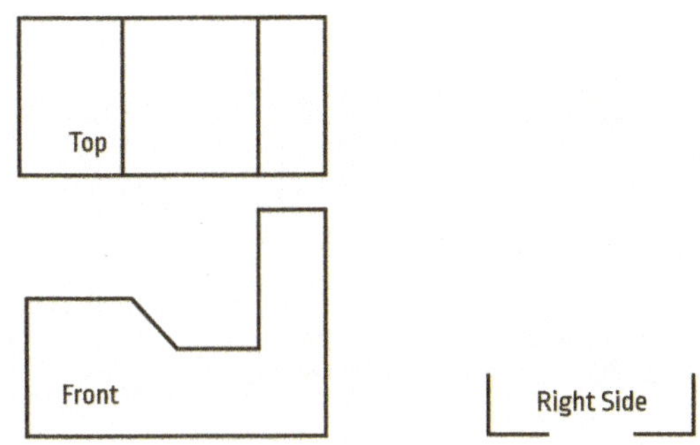

Directions: Sketch the three views using the drawing provided.

5.

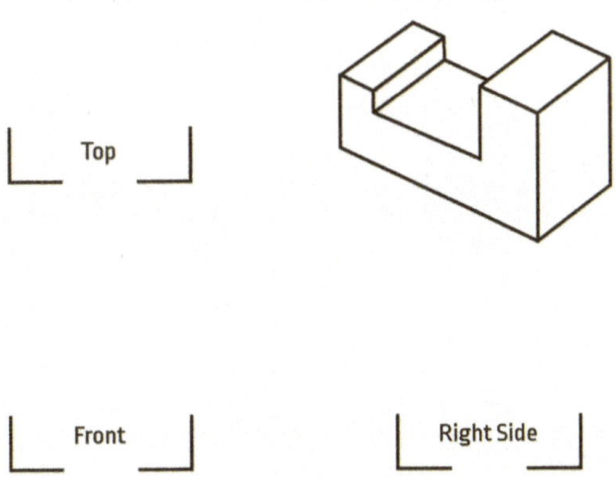

6.

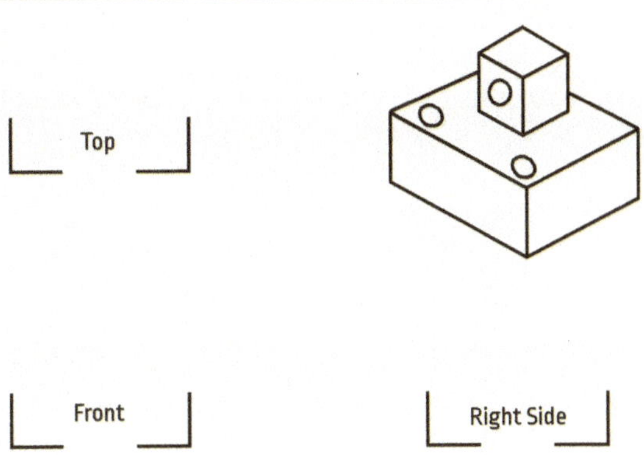

7.

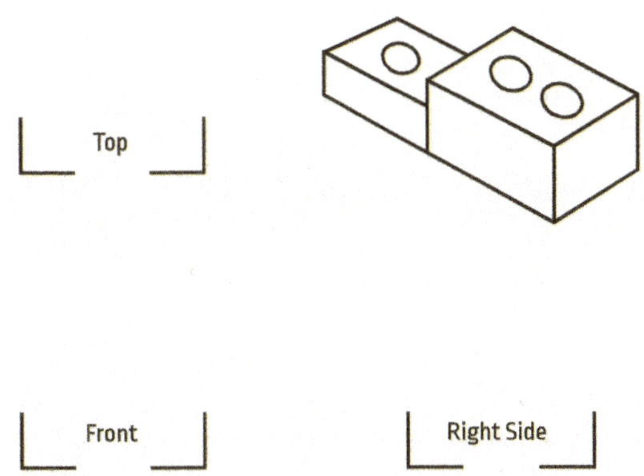

8.

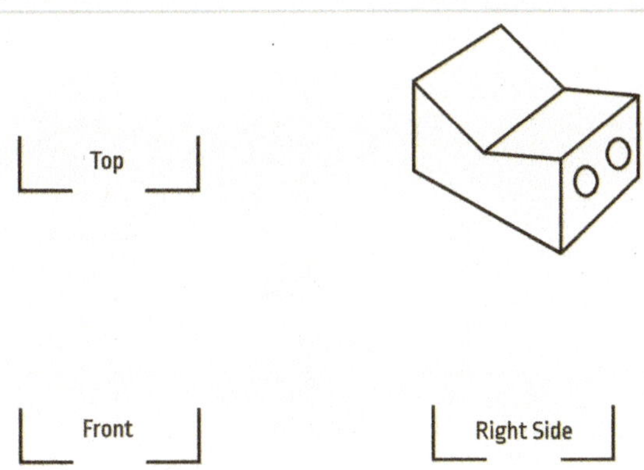

Notes

CHAPTER 2: Understanding Welding Symbols

Chapter Objective

After learning the information in this chapter you will be able to:

- Identify the parts of a welding symbol
- Determine the function of different welding symbols
- Identify fillet weld symbols
- Identify groove weld symbols
- Understand welding process symbols
- Compare weld contour symbols
- Learn nondestructive inspection symbols
- Determine dimensioning process

Key Terms

Weld symbols
Arrow side
Other side
Groove welds
Fillet welds
Complete joint penetration (CJP)
Partial joint penetration (PJP)
Process identification
Process designation
ISO standards
Dimensioning

When engineers and designers produce a drawing it can be difficult for the information to be related to the floor welder. Information can easily be missed, especially when the engineers and designers rarely get to speak to the welder.

It is sometimes difficult for engineers and designers to convey necessary information such as weld size, weld type, type of joint, how a part may be stressed, product economics and efficient manufacturing practices to the floor welder using an engineering drawing. Information can easily be missed, especially when the engineers and designers rarely get a chance to speak to the welder. I have personally worked in and visited several manufacturing plants and have seen major disconnects between the people making the prints and the welders who cannot understand the print. This can cause problems in the flow of product in a weld shop and is one of the most common reasons for re-work, distortion, poor welder morale, fatigue, defective products and higher costs.

Welding has always been a major factor in the manufacturing industry. It is vital that several pieces of critical information be indicated on the drawing so the welder can perform his or her job correctly.

To address this need, the American Welding Society (AWS) established a standard, universal set of symbols for every aspect of process and efficiency in the welding industry. The governing body for this symbol system is the AWS A2.4: 2012. Anyone who creates drawings for a welding facility should understand AWS A2.4: 2012. When designers do not have a solid understanding of the system, both the welder and the designer can become confused. The symbols used in AWS A2.4:2012 contain a great deal of information, so you might find them intimidating at first. The easiest way to learn to read these symbols is to break them down into simple steps, beginning with the most important information and working toward the more complex components. Once you understand this symbol system, you will find it easy to gather a good deal of information from a relatively small symbol.

Make Up of Welding Symbols

All welding symbols start with a reference line (Fig. 2.1). The reference line is an information baseline that is usually composed of three different parts: the reference line, the arrow and the tail. These reference lines are about 1.0 in. long and are always horizontal. Most of the needed information is placed on the reference line. The reference line will contain one welding symbol. This symbol will show the type of weld joint you are to use. The arrowhead on one end of the line points to where the weld is to be placed in the weld joint. The tail bracket can contain additional information such as the welding process to be used, special details, or notes found elsewhere on the drawing.

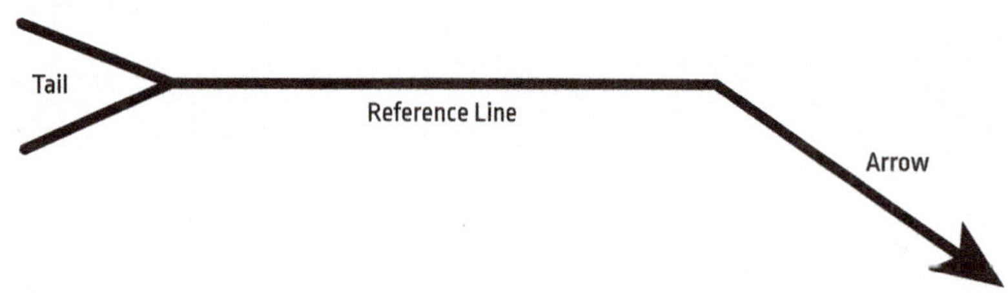

Figure 2.1 All welding symbols begin with a reference line, an arrow pointing to the weld and a tail bracket.

A welding symbol consists of up to eight different components to show the weld placement, type of weld, size of weld or length of the weld. The information contained in the welding symbol is easy to learn and accurate, which eliminates the need for written directions or notes. The elements contained in the welding symbol are the reference line, arrow, weld symbol, supplementary symbols, weld dimensions, finishing symbols, tail and welding process (Fig. 2.2). In contrast, a weld symbol indicates the type of weld that should be placed in the weld joint.

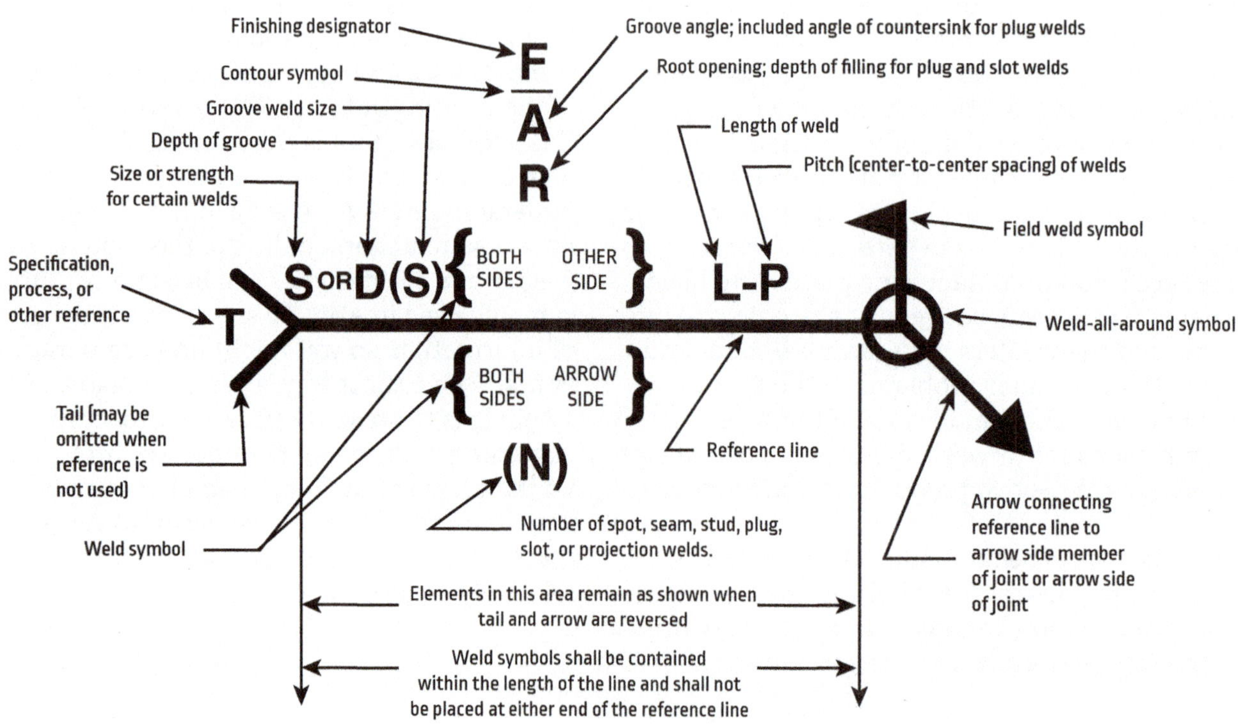

Figure 2.2

The arrow can point in different directions but must ultimately point to the area to be welded. The weld symbol may be placed on either the top or the bottom of the reference line. The top of the reference line is always considered the *other side*, and the bottom of the reference line is considered the *arrow side*. This does not change, no matter which way the arrow points to the joint.

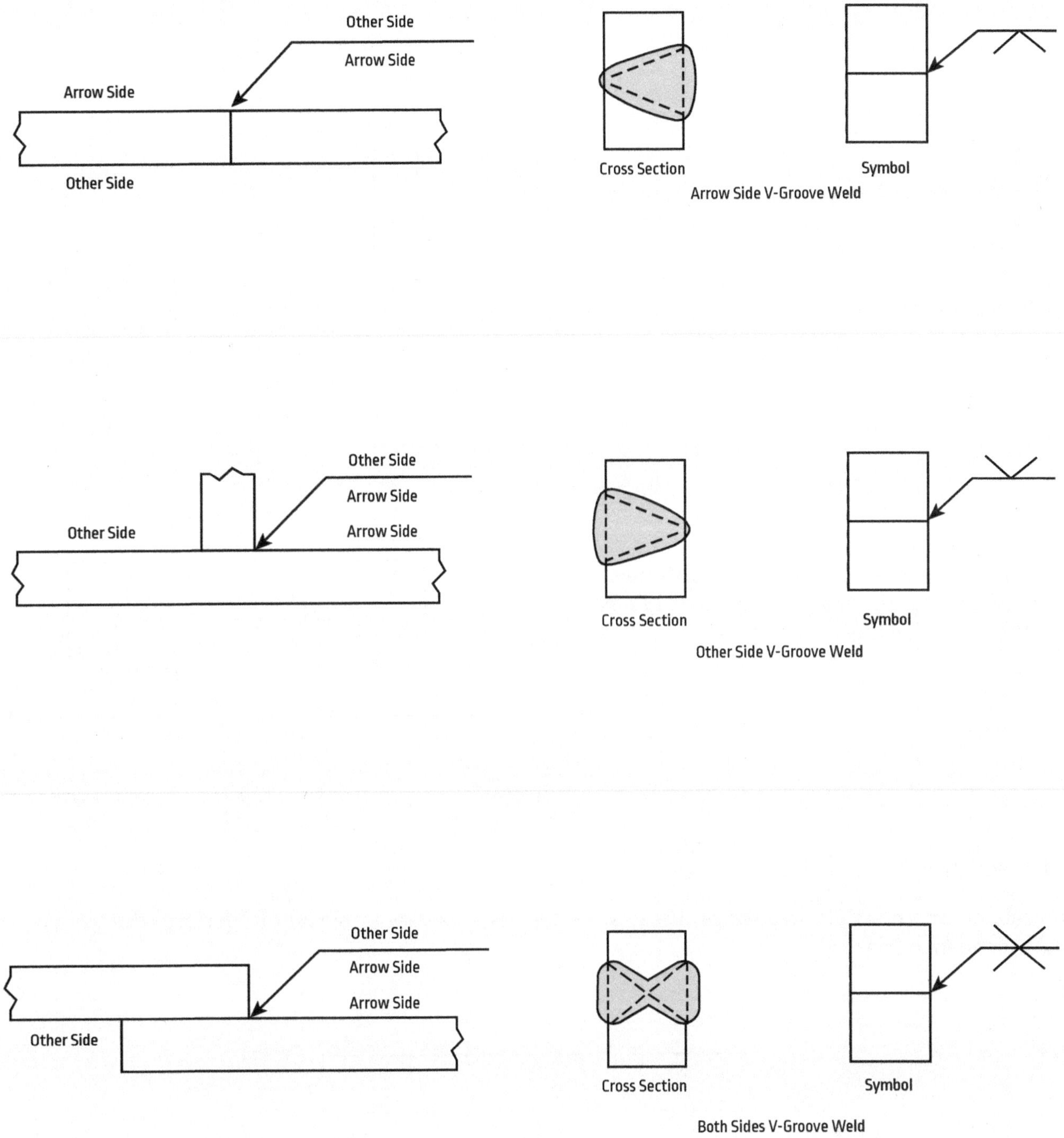

Figure 2.3 In the examples above, we can see how the weld symbols place the weld in the different joints.

Arrow Side

The arrow side of the welding symbol places the weld on the nearest or same side of the joint that the arrow is pointing to. The arrow side indicates that the weld is to be placed exactly where the arrow points (Fig. 2.4a).

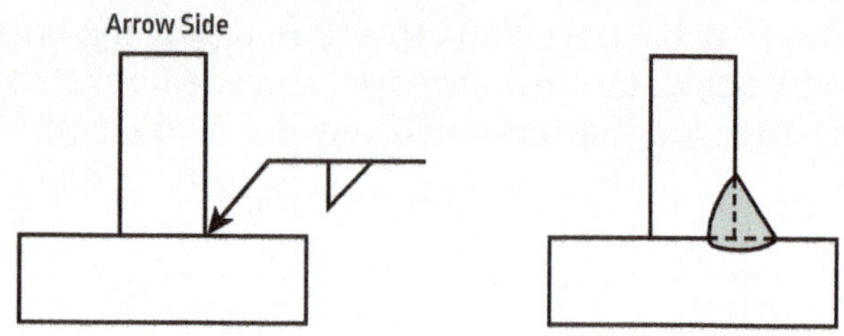

Figure 2.4a The fillet weld symbol is on the arrow side, indicating that the weld is placed exactly where the arrow points.

Other Side

The other side places the weld on the opposite side of the weld joint. The other side is the top of the reference line and shows that the weld is to be placed on the opposite or furthest side from where the arrow is pointing (Fig. 2.4b).

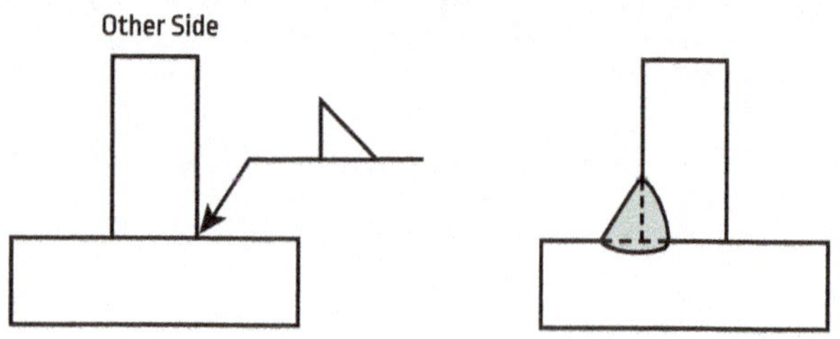

Figure 2.4b The fillet weld symbol is placed on the other side of the reference line. This shows the weld being placed opposite to where the arrow is pointing.

Both Sides

A part that is to be welded on both sides of the weld joint is indicated by the weld symbol placed on both sides of the reference line (Fig. 2.4c).

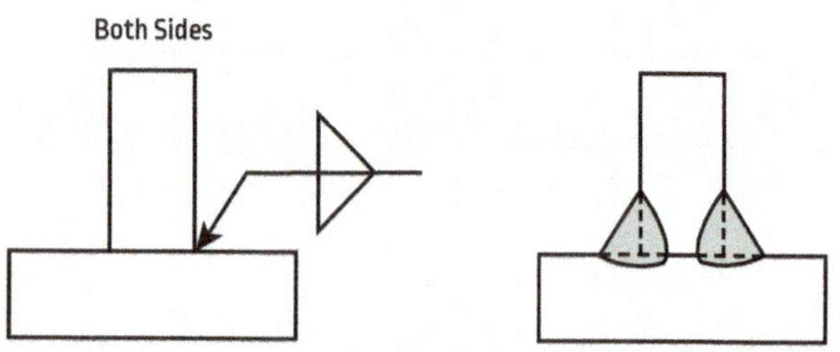

Figure 2.4c The fillet welds are placed directly across from each other on the reference line. This indicates both sides receive the weld.

Multiple Reference Lines

Many welds require multiple steps to get the desired results. It is very important to perform the weld in a particular sequence (Fig. 2.5).

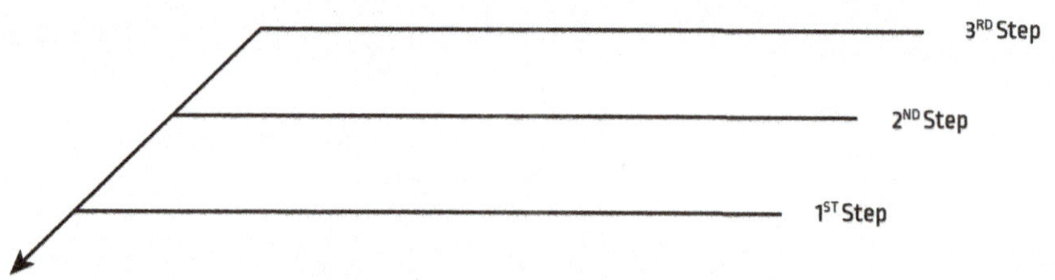

Figure 2.5 When multiple reference lines are used, the first operation is always closest to the arrow.

We have established how a basic welding symbol is created, which side is the arrow side and which side is the other side, and the significance of the placement of the welding symbol. Now we can begin to add additional information to the symbol. The AWS has different definitions for the terms weld symbol and welding symbol. A weld symbol is a symbol that is represented by one of the universal symbols seen in Tables 2.1 through 2.3. Weld symbols often look similar to the weld joint to be used. It is important to review and learn the different weld symbols.

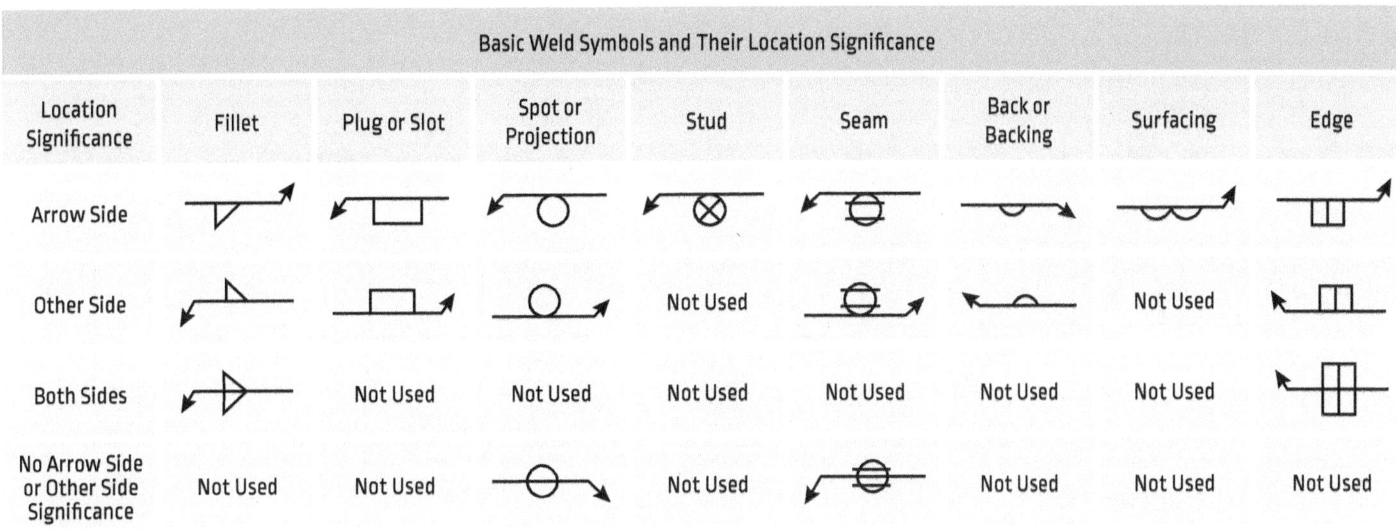

Basic Weld Symbols and Their Location Significance								
Location Significance	Fillet	Plug or Slot	Spot or Projection	Stud	Seam	Back or Backing	Surfacing	Edge
Arrow Side								
Other Side				Not Used			Not Used	
Both Sides		Not Used	Not Used	Not Used	Not Used	Not Used	Not Used	
No Arrow Side or Other Side Significance	Not Used	Not Used		Not Used		Not Used	Not Used	Not Used

Table 2.1

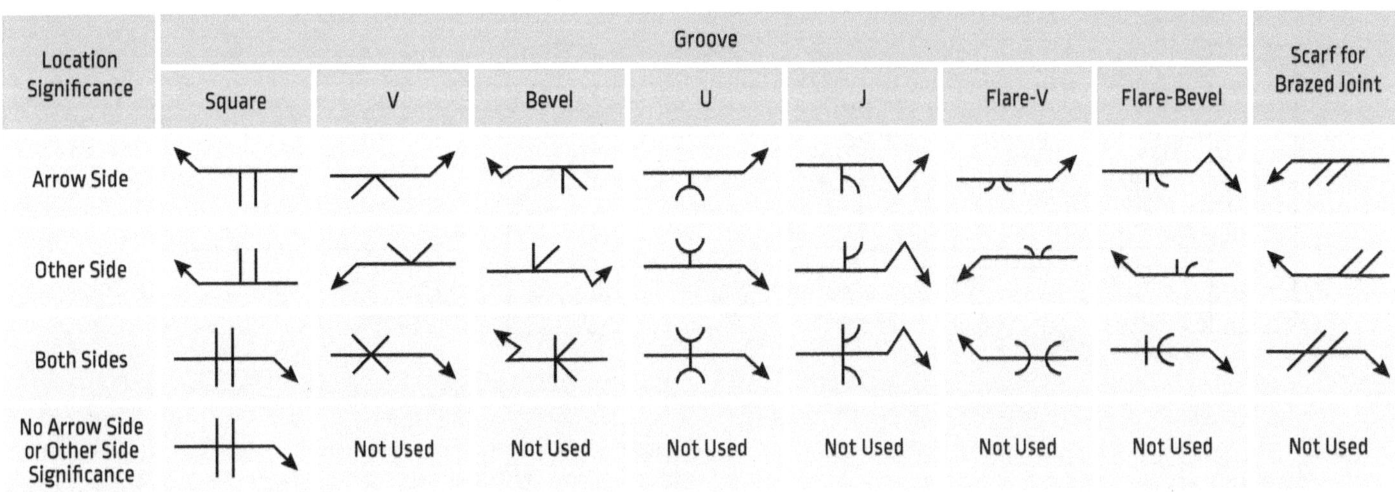

Location Significance	Groove							Scarf for Brazed Joint
	Square	V	Bevel	U	J	Flare-V	Flare-Bevel	
Arrow Side								
Other Side								
Both Sides								
No Arrow Side or Other Side Significance		Not Used	Not Used	Not Used	Not Used	Not Used	Not Used	Not Used

Table 2.2

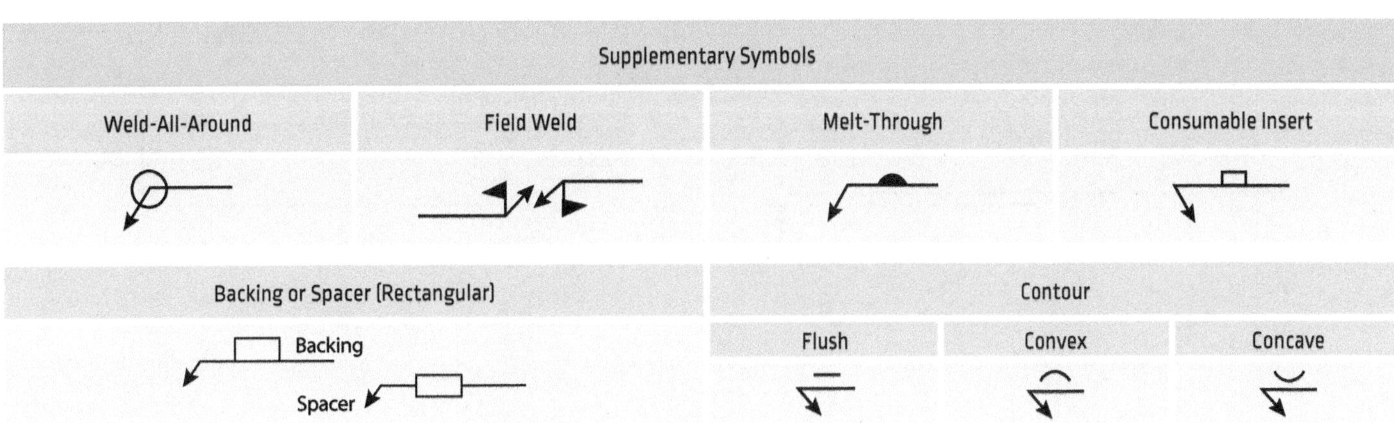

Supplementary Symbols				
Weld-All-Around	Field Weld	Melt-Through	Consumable Insert	
Backing or Spacer (Rectangular)	Contour			
	Flush	Convex	Concave	
Backing / Spacer				

Table 2.3

Groove Weld Symbols

Groove weld symbols come in several different variations and combinations. The main types of groove welds are square, bevel, V, U, J, flare-bevel and flare-V grooves. A weld joint can have one type of groove on one side and a different type of groove on the other side. Groove welds should be specified as complete joint penetration (CJP) or partial joint penetration (PJP).

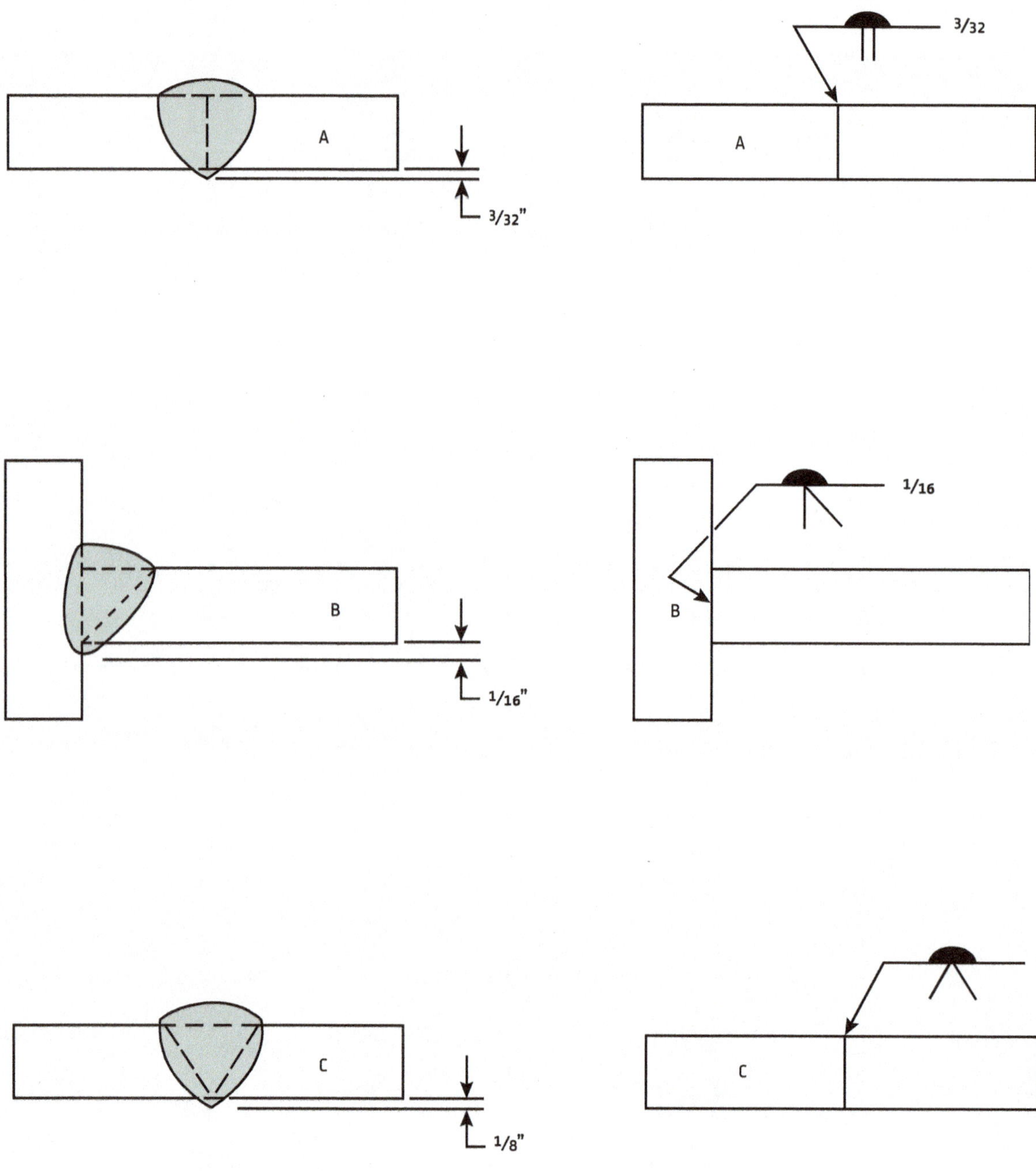

Figure 2.6 Complete joint penetration welds are indicated on the weld symbol with a "solid" half circle, also referred to as "melt-through".

CJP (Complete Joint Penetration)

An example of a CJP joint would be a butt joint with a V-groove. This weld requires complete joint penetration all the way through to the other side of the weld joint. The main factor that determines whether CJP or PJP is required is usually the thickness of the material, but the welding process can also have a great deal of influence on your selecting CJP or PJP. Another factor to consider is the strength of the finished part. Engineers perform painstaking calculations to determine whether a weld joint needs complete penetration or partial penetration to meet design and strength requirements. When we butt two pieces of material together, we usually need a CJP weld to ensure the necessary strength the part is specified for. If the component does not have certain strength requirements, then a CJP weld is not necessary. A PJP weld would be adequate.

Complete joint penetration is necessary on most butt joints. Depending on the thickness of the material and the welding process this can be done with single or multipass welds. The symbol indicates a 70° included bevel angle with complete penetration.

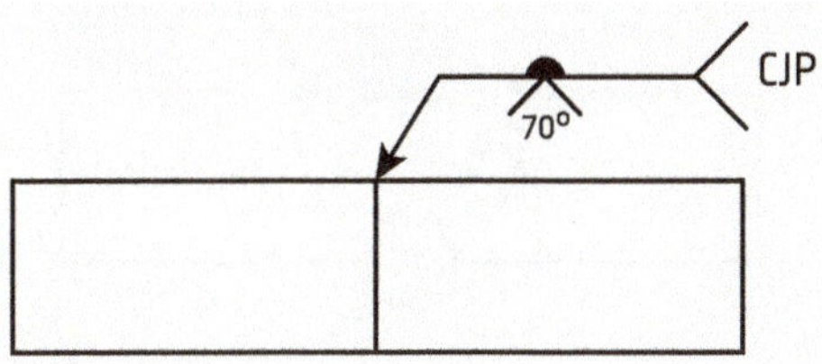

Figure 2.7

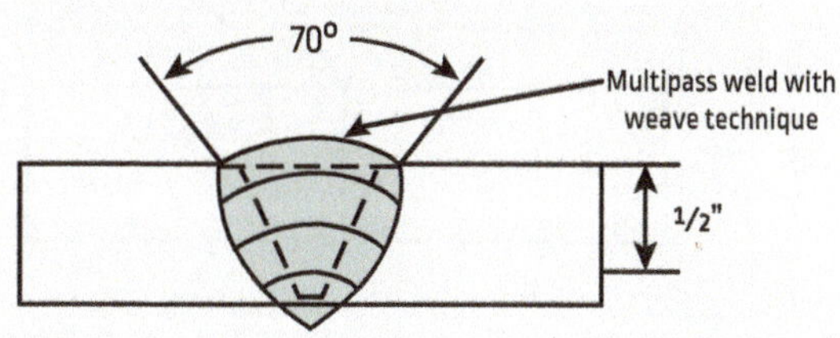

Figure 2.8

When material thickness is reduced a single pass weld can often be used. Many times a "root opening" is required to obtain full penetration. The symbol indicates a 1/8 in. root opening with complete joint penetration.

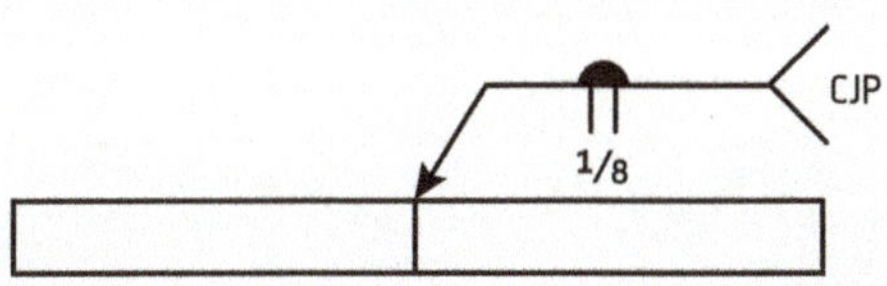

Figure 2.9

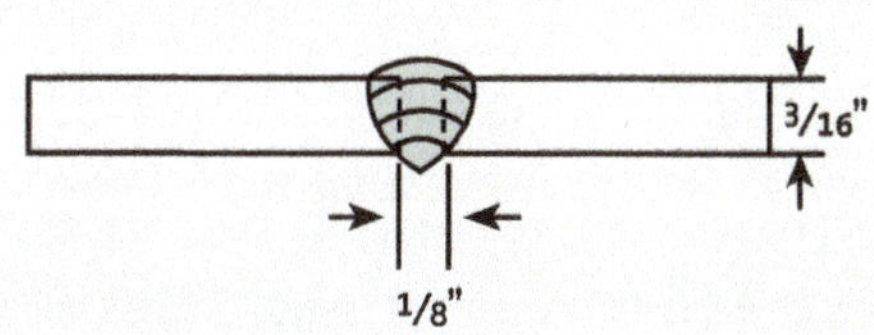

Figure 2.10

With thicker materials a double V-groove joint is used for a complete penetration weld. Depending on process and material thickness, this joint type can be much more economical than a single V-groove joint.

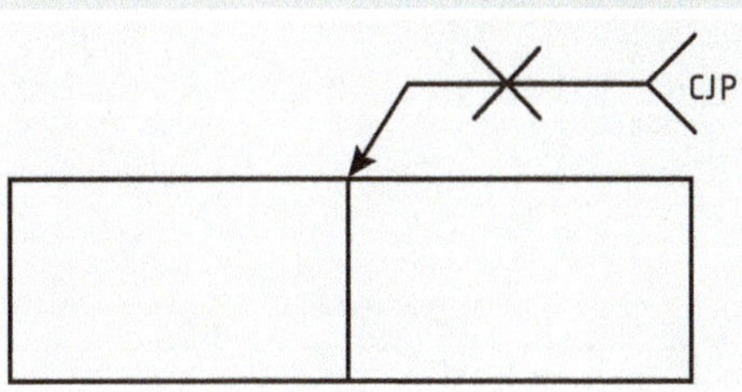

Figure 2.11

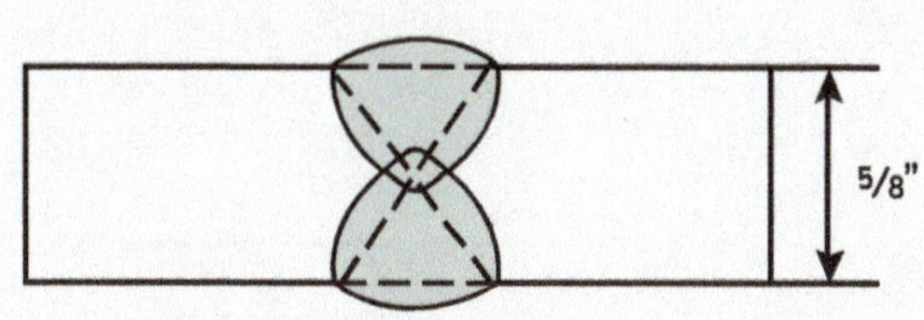

Figure 2.12

When using a double V-groove the depth of the groove and the depth of the weld are both indicated to ensure complete penetration.

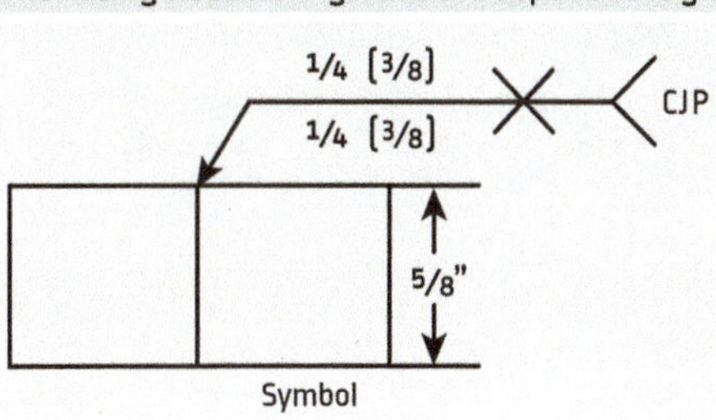

Figure 2.13

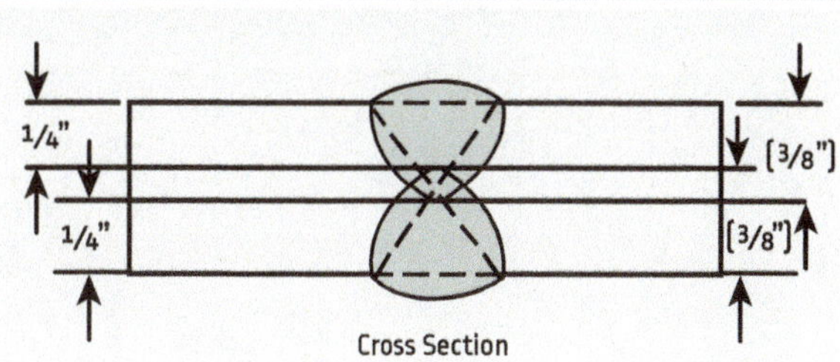

Figure 2.14

Another common way to ensure complete penetration is to use "back" welds. Back welds are done after the main weld is completed. Back welds fuse into the main weld ensuring complete penetration. The bottom of the main weld is prepped by grinding, carbon arc gouging or other means before the back weld is applied.

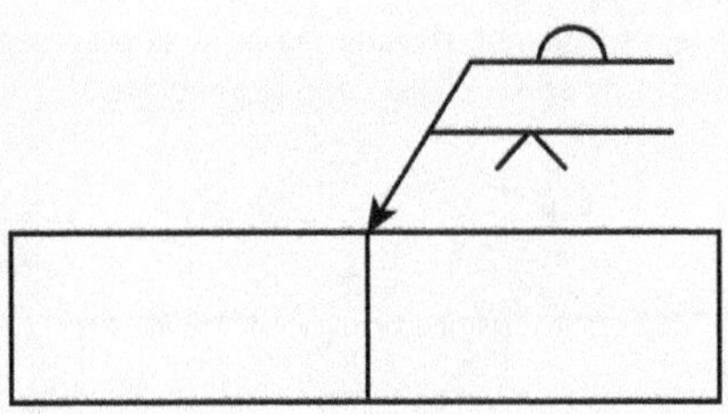

Figure 2.15

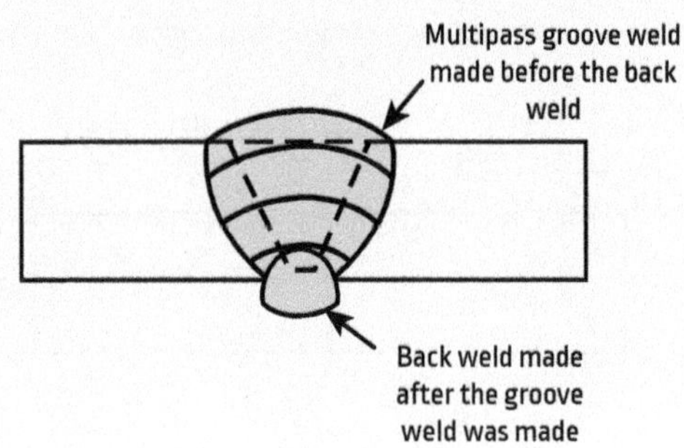

Figure 2.16

"Backing" welds differ from "back welds" in the simple fact that they are performed before the main weld, not after as with the back weld. A backing weld helps keep the main weld from falling through the bottom of the material. The only difference between the two is the order of operations.

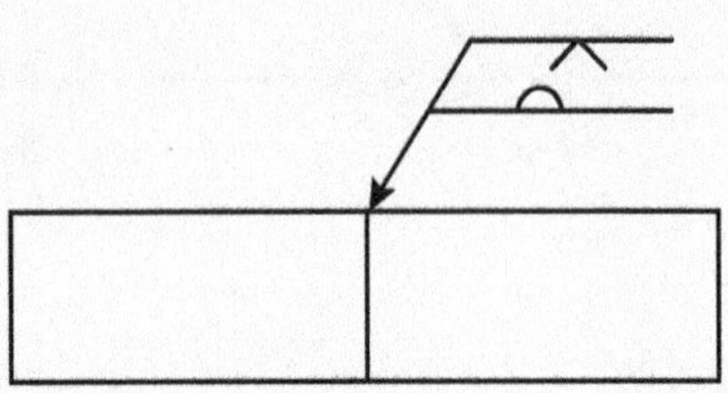

Figure 2.17

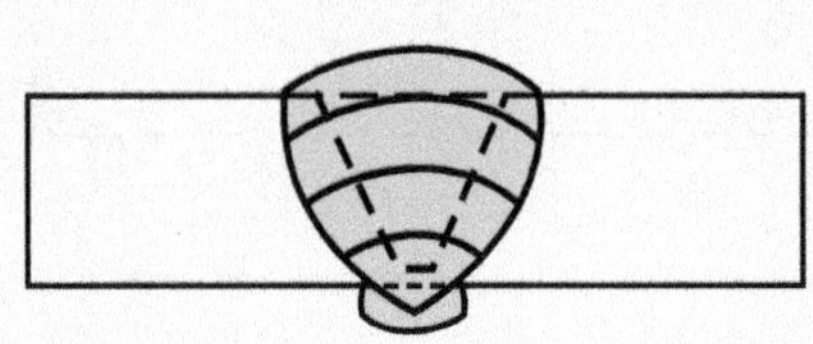

Figure 2.18

PJP (Partial Joint Penetration)

PJP welds are generally used when CJP welds are not necessary. PJP welds are much faster and easier to prepare. They also reduce distortion and weld metal volume, which generally equals cost reduction. PJP welds are used in several different types of grooves. Some butt joints can be completed using PJP.

Whether to use CJP or PJP is not a decision for the welder; rather, it will be determined by the designer using engineered calculations. The welder should simply look at the tail section of the welding symbol, where the type of penetration required will be indicated with the letters CJP or PJP. The welder can then perform the weld according to the information on the print.

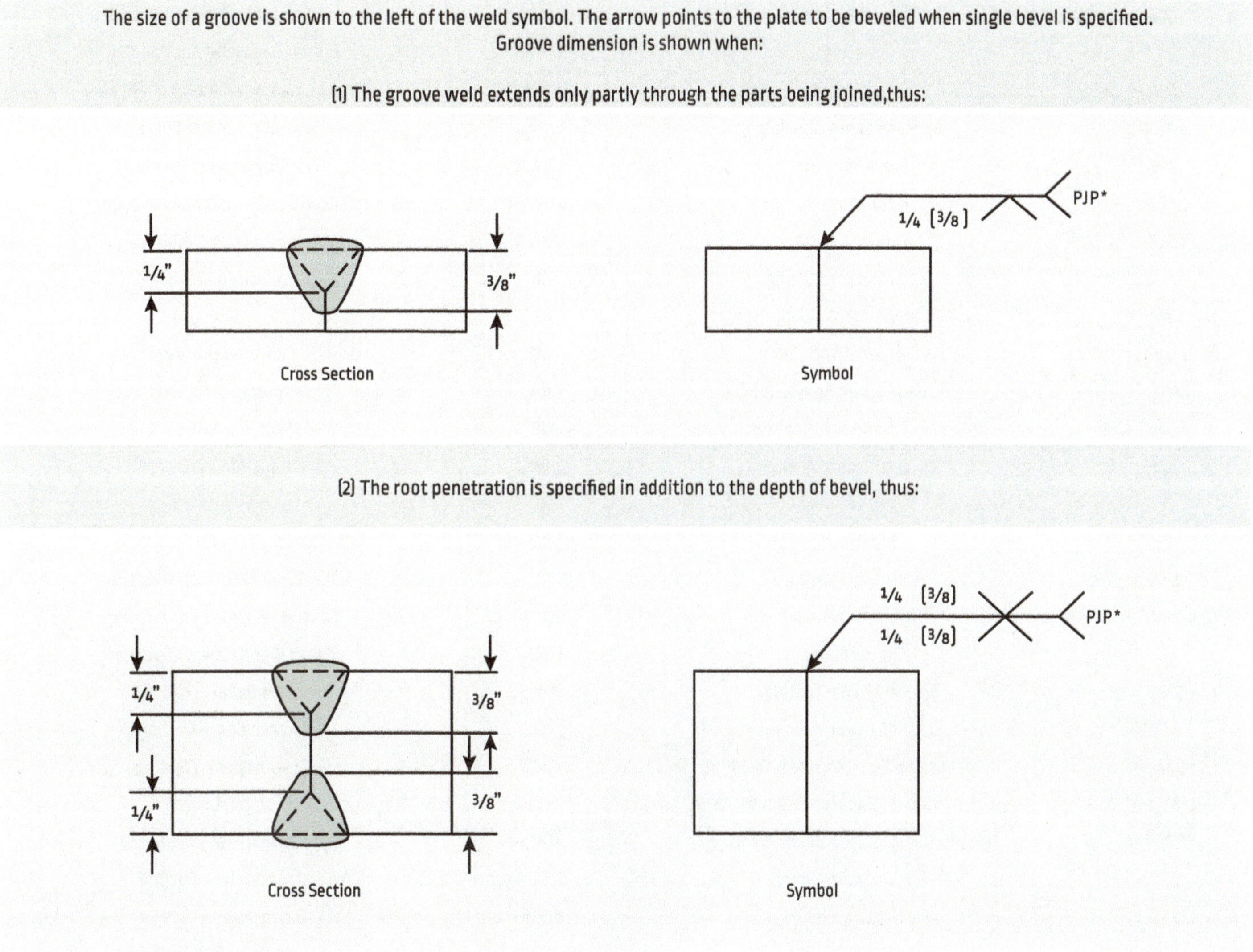

Figure 2.19 Partial joint penetration is indicated in the tail of the weld symbol.

Welding Process Identification Symbols

We are all somewhat familiar with the four most well-known and widely used welding and cutting processes: shielded metal arc welding (SMAW), gas metal arc welding (GMAW or MIG), flux-cored arc welding (FCAW) and gas tungsten arc welding (GTAW or TIG). What many people don't realize is that the AWS recognizes well over 100 different welding processes, each with its own letter designation.

LETTER DESIGNATIONS OF WELDING AND ALLIED PROCESSES			
Letter Designation	Processes and Variations	Letter Designation	Processes and Variations
AAW	Air Acetylene Welding	FOC	Flux Cutting
ABW	Arc Braze Welding	FOW	Forge Welding
AC	Arc Cutting	FRW	Friction Welding
AHW	Atomic Hydrogen Welding	FS	Furnace Soldering
AOC	Oxygen Arc Cutting	FW	Flash Welding
ASP	Arc Spraying	GMAC	Gas Metal Arc Cutting
AW	Arc Welding	GMAW	Gas Metal Arc Welding
B	Brazing	GMAW-P	Pulsed Gas Metal Arc Welding
BB	Block Brazing	GMAW-S	Short Circuit Gas Metal Arc Welding
BMAW	Bare Metal Arc Welding	GTAC	Gas Tungsten Arc Cutting
BW	Braze Welding	GTAW	Gas Tungsten Arc Welding
CABW	Carbon Arc Braze Welding	GTAWP	Pulsed Gas Tungsten Arc Welding
CAC	Carbon Arc Cutting	HPW	Hot Pressure Welding
CAC-A	Air Carbon Arc Cutting	IB	Induction Brazing
CAW	Carbon Arc Welding	INS	Iron Soldering
CAW-G	Gas Carbon Arc Welding	IRB	Infrared Brazing
CAW-S	Shielded Carbon Arc Welding	IRS	Infrared Soldering
CAW-T	Twin Carbon Arc Welding	IS	Induction Soldering
CEW	Co-extrusion Welding	IW	Induction Welding
CW	Cold Welding	LBC	Laser Beam Cutting
DB	Dip Brazing	LBC-A	Laser Beam Air Cutting
DFB	Diffusion Brazing	LBC-EV	Laser Beam Evaporative Cutting
DFW	Diffusion Welding	LBC-IG	Laser Beam Inert Gas Cutting
DS	Dip Soldering	LBC-O	Laser Beam Oxygen Cutting
EBC	Electron Beam Cutting	LBW	Laser Beam Welding
EBW	Electron Beam Welding	LOC	Oxygen Lance Cutting
EBW-HV	High Vacuum Electron Beam Welding	OAW	Oxyacetylene Welding
EBW-MV	Medium Vacuum Electron Beam Welding	OC	Oxygen Cutting
EBW-NV	Nonvacuum Electron Beam Welding	OFC	Oxyfuel Gas Cutting
EGW	Electrogas Welding	OFC-A	Oxyacetylene Cutting
ESW	Electroslag Welding	OFC-H	Oxyhydrogen Cutting
EXB	Exothermic Brazing	OFC-N	Oxynatural Gas Cutting
EXBW	Exothermic Braze Welding	OFC-P	Oxypropane Cutting
EXW	Explosion Welding	OFW	Oxyfuel Gas Welding
FB	Furnace Brazing	OHW	Oxyhydrogen Welding
FCAW	Flux Cored Arc Welding	PAC	Plasma Arc Cutting
FCAW-G	Gas Shielded Flux Cored Arc Welding	PAW	Plasma Arc Welding
FCAW-S	Self-shielded Flux Cored Arc Welding	PEW	Percussion Welding
FLB	Flow Brazing	PGW	Pressure Gas Welding
FLOW	Flow Welding	POC	Metal Powder Cutting
FLSP	Flame Spraying	PSP	Plasma Spraying

Table 2.4 We can see that there are many welding processes other than the four most common.

LETTER DESIGNATIONS OF WELDING AND ALLIED PROCESSES			
Letter Designation	Processes and Variations	Letter Designation	Processes and Variations
PW	Projection Welding	SSW	Solid-state Welding
RB	Resistance Brazing	SW	Arc Stud Welding
ROW	Roll Welding	TB	Torch Brazing
RS	Resistance Soldering	TC	Thermal Cutting
RSEW	Resistance Seam Welding	TCAB	Twin Carbon Arc Brazing
RSEW-HF	High-frequency Seam Welding	THSP	Thermal Spraying
RSEW-I	Induction Seam Welding	TS	Torch Soldering
RSW	Resistance Spot Welding	TW	Thermite Welding
RW	Resistance Welding	USS	Ultrasonic Soldering
S	Soldering	USW	Ultrasonic Welding
SAW	Submerged Arc Welding	UW	Upset Welding
SAW-S	Series Submerged Arc Welding	UW-HF	High-frequency Upset Welding
SMAC	Shielded Metal Arc Cutting	UW-I	Induction Upset Welding
SMAW	Shielded Metal Arc Welding	WS	Wave Soldering

Table 2.4 Continued

Welding Process Designations

Most of the welding processes in Table 2.4 are used in very specialized applications. In this book, we will focus on the most common welding processes that most of us have probably used or at least heard of. A list of these processes and their letter designations is shown in Table 2.5.

SMAW	Shielded Metal Arc Welding (STICK)
GMAW	Gas Metal Arc Welding (MIG), Spray Transfer Is Assumed Unless -P Or -S Is Specified
GMAW-P	Pulsed Gas Metal Arc Welding
GMAW-S	Short Circuit Gas Metal Arc Welding
GTAW	Gas Tungsten Arc Welding (TIG)
FCAW	Flux Cored Arc Welding, With Or Without Gas Unless
FCAW-G	Gas Shielded Flux Cored Arc Welding
FCAW-S	Self Shielded Flux Cored Arc Welding
PAW	Plasma Arc Welding
SAW	Submergerd Arc Welding
ESW	Electroslag Welding
LBW	Laser Beam Welding
EBW	Electron Beam Welding
PW	Projection Welding
RSW	Resistance Spot Welding
RSEW	Resistance Seam Welding

Table 2.5

Additional Process Designations Symbols

As you review Tables 2.4 and 2.5, keep in mind that several of the welding processes can be automated to varying extents. For example, GMAW is considered a semi-automatic welding process when it is performed by a welder using an automatic wire feeder because the machine automatically provides a smooth and consistent feeding of the filler metal. The process is not considered fully automated because the welder physically engages the machine, moves the welding gun across the joint and controls travel speeds, arc lengths and gun angles. If the welder's job is done using automation, then the process is considered automatic. If the welding gun is transferred to a robot arm, the nature of the process changes again. See Table 2.6 for the list of additional process identifications.

Symbol	Method
MA	Manual
SA	Semi-Automatic
ME	Machine
AU	Automatic

Table 2.6

Weld Finishing Method Symbols

Many times, the welds we produce need a final finishing step. This can be done using several different methods, which are identified in Table 2.7. Sometimes the finishing step is needed to reduce stress in the part. For example, another part may have to fit in a particular position, which can't be accomplished without grinding the weld down a little bit. Other times, the finishing step may only be required for painting or coating purposes.

	Symbol	Method
1.	C	Chipping
2.	G	Grinding
3.	H	Hammering
4.	M	Machining
5.	R	Rolling
6.	P	Peening
7.	U	Unspecified

Table 2.7

Non-Destructive Testing Designation Symbols

After we have followed all the directions on the print and made a completed weld using all of the important information available to us, our finished weld must usually undergo a quality assurance inspection process. Weld inspection methods can range from a simple visual inspection all the way to a radiographic (X-ray) test. Non-destructive testing (NDT) is done using several different methods (Table 2.8). The method used varies depending on the situation. The important thing to remember is that NDT does not destroy the weld. The NDT inspection method will be identified on the print and must be followed.

Symbol	Method
VT	Visual Testing
MT	Magnetic Particle Testing
PT	Penetrant Testing
UT	Ultrasonic Testing
RT	Radiographic Testing
ET	Electromagnetic (Eddy Current) Testing
LT	Leak Testing
NRT	Neutron Radiographic Testing
AET	Acoustic Emission Testing
PRT	Proof Testing

Table 2.8

PRINT	COMMENTS
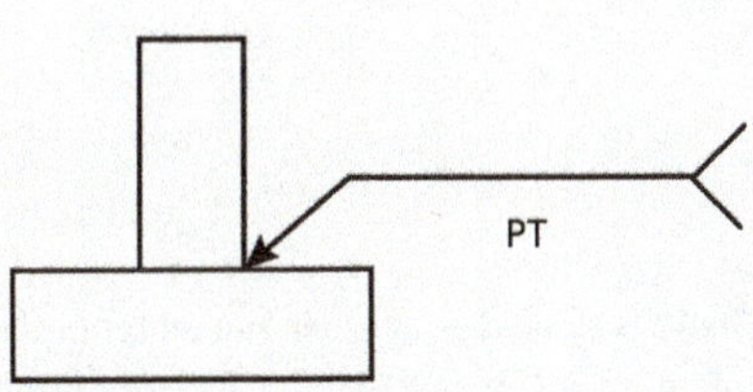	The symbol indicates a dye penetrant test on the arrow side of the weld joint.

Figure 2.20

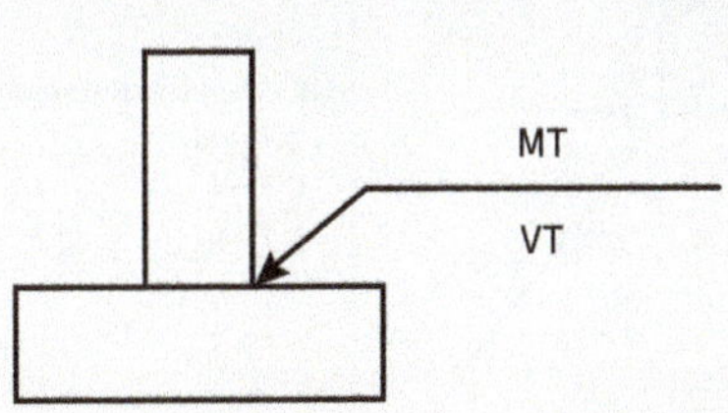	The symbol indicates a visual test on the arrow side and a magnetic particle test on the other side.

Figure 2.21 Non-destructive testing is a critical component in most quality control systems.

Non-Destructive Examination Symbols

	PRINT	COMMENTS
1.	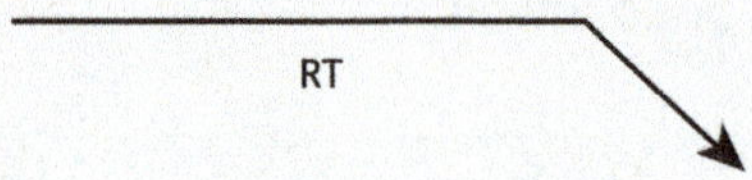	The symbol indicates a radiograph is to be performed on the arrow side of the weld
2.	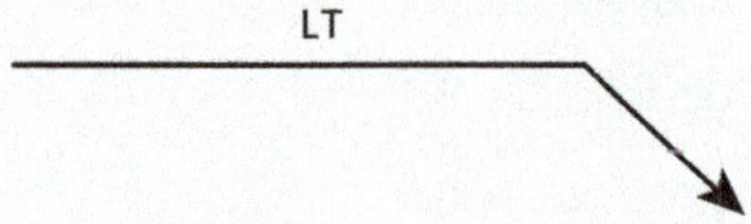	The symbol indicates a leak test is to be performed on the other side of the weld.
3.	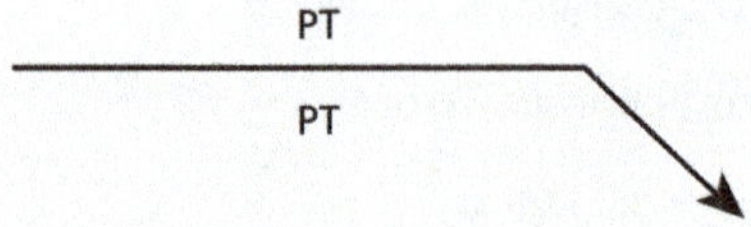	The symbol indicates a dye penetrant test is to be performed on both sides of the weld.
4.	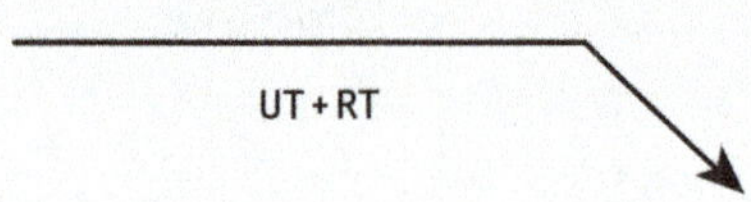	The symbol indicates an ultrasonic test in addition to a radiograph test. The "+" sign is used to indicate multiple examinations. This can be done on the arrow side, other side or no significance.
5.	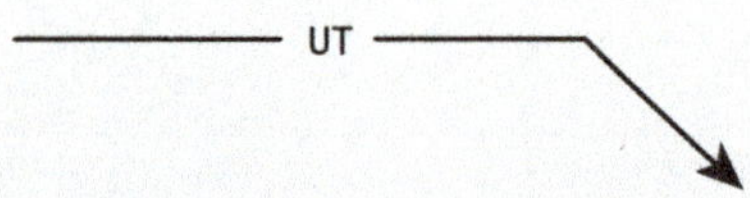	When the symbol is placed in the center of the line, it has no arrow side or other side significance.

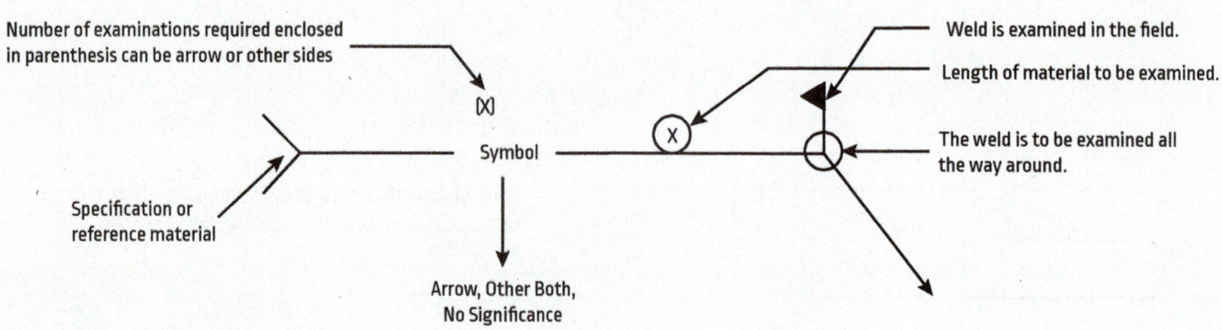

Table 2.9 The illustrations show that there can be more than one type of NDT used on the same weldment.

ISO International Standard Welding Symbols

The AWS welding symbols are probably the most commonly used symbols in the United States, but there is another set of welding symbols with which we should familiarize ourselves: the International Organization for Standardization (ISO) symbols (Table 2.10). I have seen ISO symbols being used more and more in recent years. The International Standard 2553 is a publication of welding symbols used and approved by the ISO to ensure consistency on a world-wide basis. These symbols are somewhat different from the AWS system, but you will be able to pick them up quickly with a little effort. The ISO system is becoming more prevalent, especially in plants that ship their product to other countries or construct products for other countries. The shop you work in will determine which of the two equally effective systems you will use.

No.	Designation	Symbol
1.	Butt Weld Between Plates With Raised Edges[1]	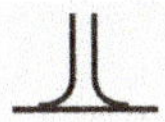
2.	Square Butt Weld	
3.	Single-V Butt Weld	
4.	Single-bevel Butt Weld	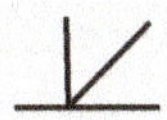
5.	Single-V Butt Weld With Broad Root Face	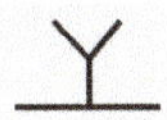
6.	Single-bevel Butt Weld With Broad Root Face	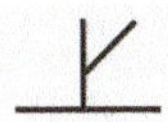
7.	Single-U Butt Weld (Parallel Or Sloping Sides)	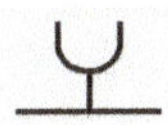
8.	Single-J Butt Weld	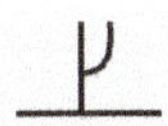
9.	Backing Run; Back Or Backing Weld/USA	
10.	Fillet Weld	
11.	Plug Weld	
12.	Spot Weld	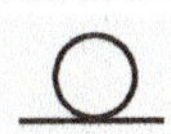
13.	Seam Weld	
14.	Steep-flanked Single-V Butt Weld	
15.	Steep-flanked Single-bevel Butt Weld	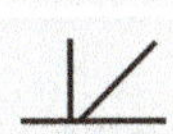
16.	Edge Weld	
17.	Surfacing Weld	
18.	Surface Joint	
19.	Inclined Joint	
20.	Fold Joint	

Table 2.10 Arrow side and other side symbols are the same for an ISO or AWS system. The two systems are very similar.

Dimensioning

Welding symbols show not only what kind of weld is needed and where the weld must be placed, but also the dimensioning of the weld. When we talk about weld dimensioning, we are referring not to the dimensions of the part but rather the measurements of the actual weld. Dimensioning of a welding symbol can also include measurements such as the legs of the weld, center to center spacing (called the pitch), the actual length of the weld, groove angles, root openings on butt welds, the size of spot welds, and angles of the countersink for plug and slot welds. Some welding symbols will show a weld on both sides of a joint. These welds may be the same size or different sizes. The size of the weld will be indicated by the welding symbol.

Size of Unequal Leg Fillet Weld

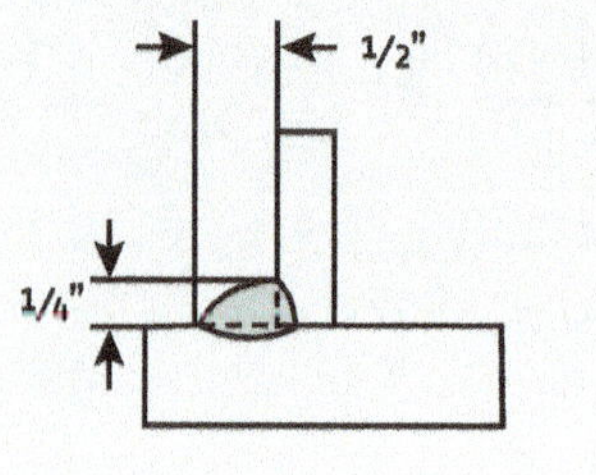

Figure 2.22

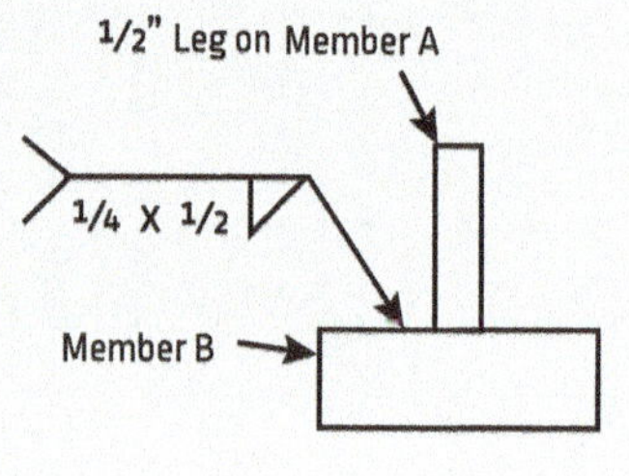

Figure 2.23

Size of Unequal Double-Fillet Welds

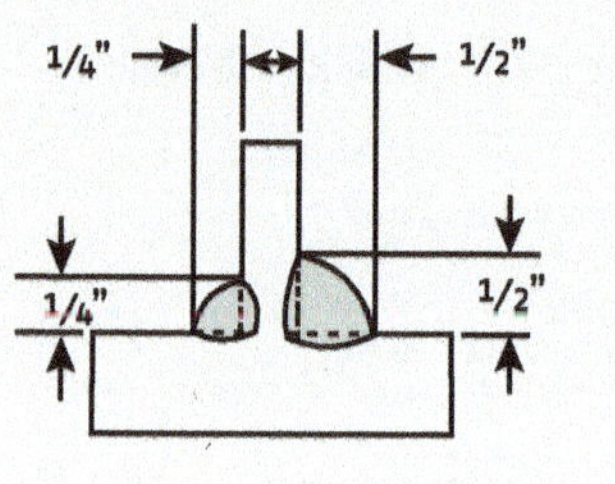

Figure 2.24

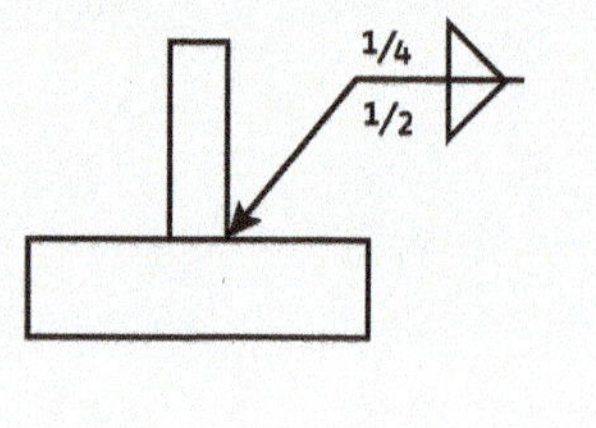

Figure 2.25

The size of the fillet weld is determined by the length of its shortest leg.

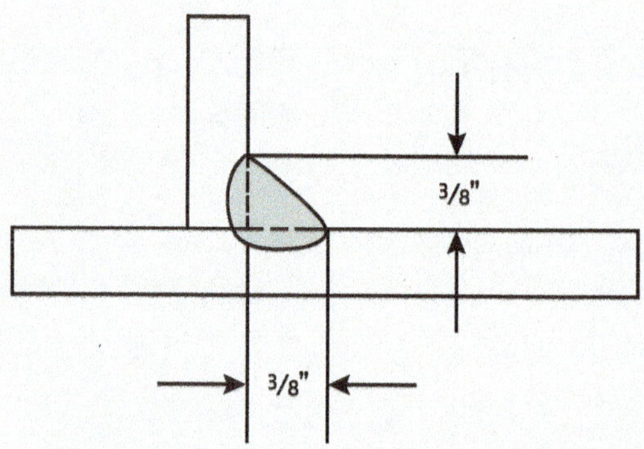

Figure 2.26

Dimensioning includes more than just weld dimensions. Other important measurements are needed to complete a specific weld.

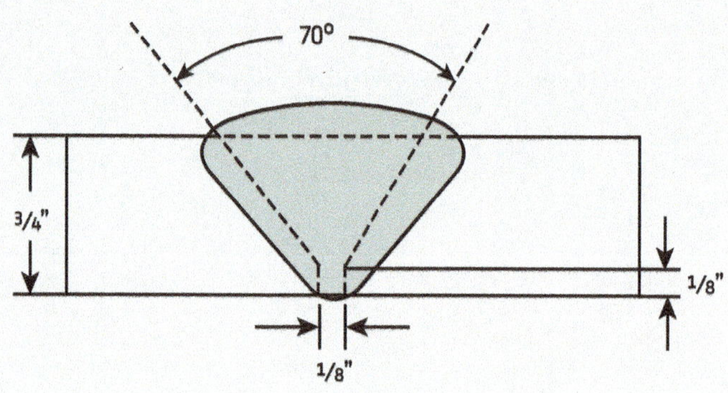

Figure 2.27

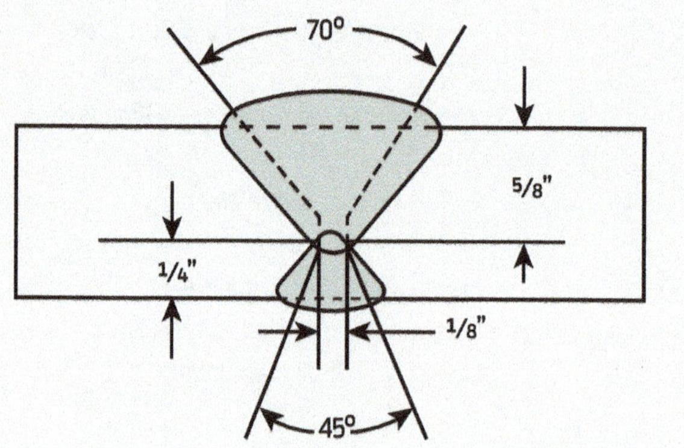

Figure 2.28

Review Exercises

Draw the correct weld as indicated by the weld symbol.

1.

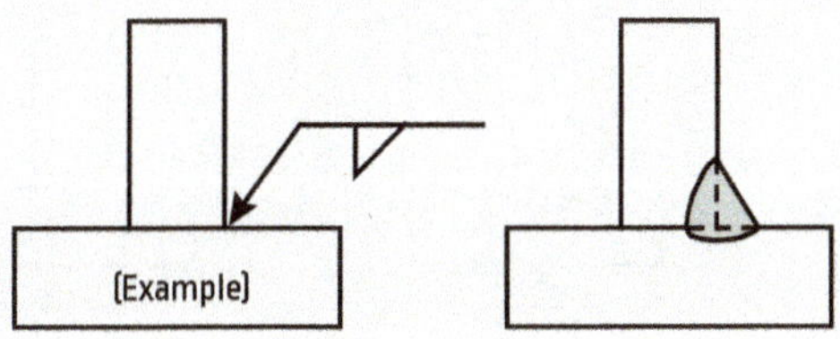

2.

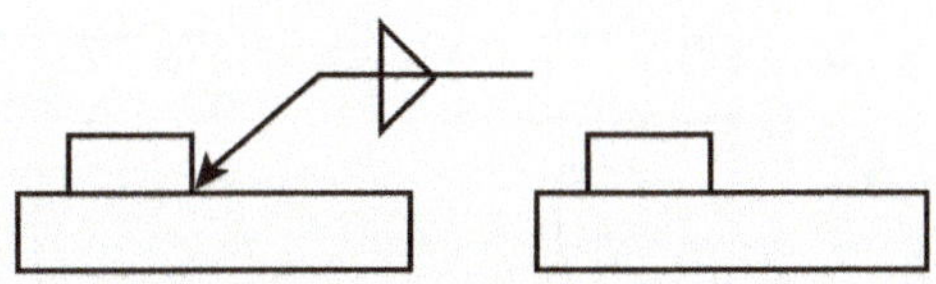

3.

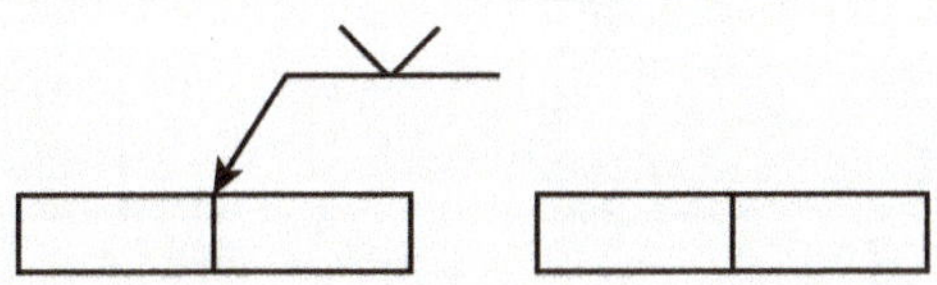

4.

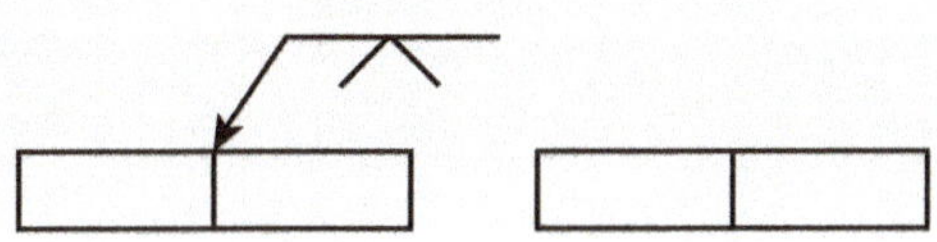

5.

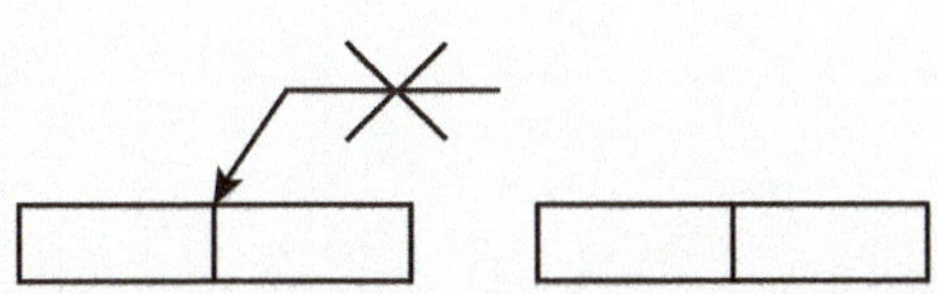

6.

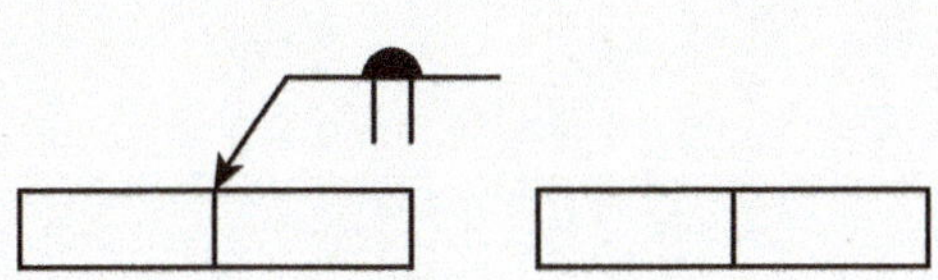

7.

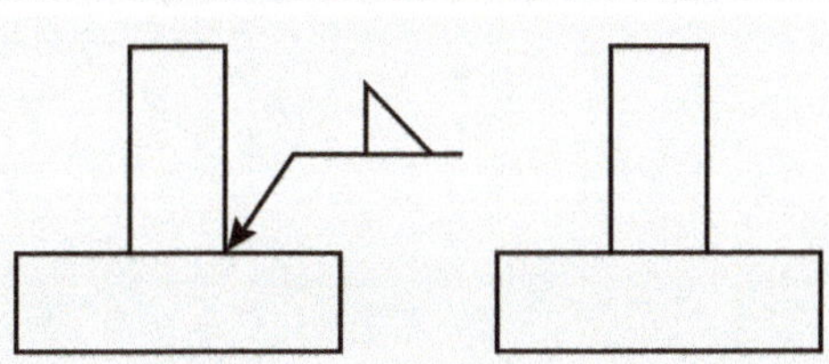

8.

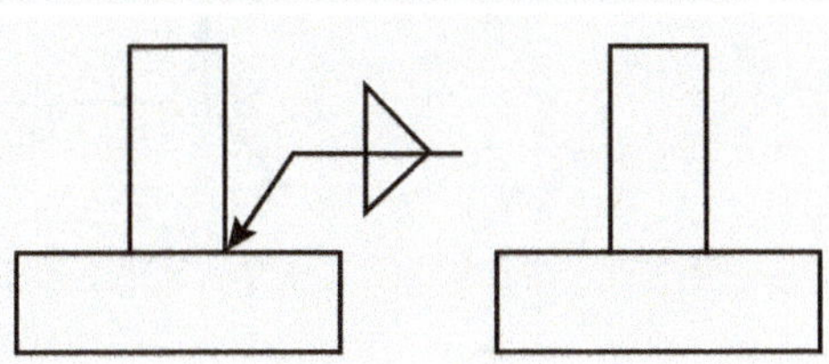

9.

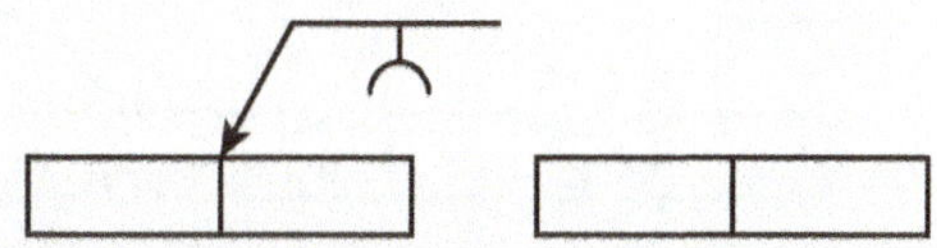

10.

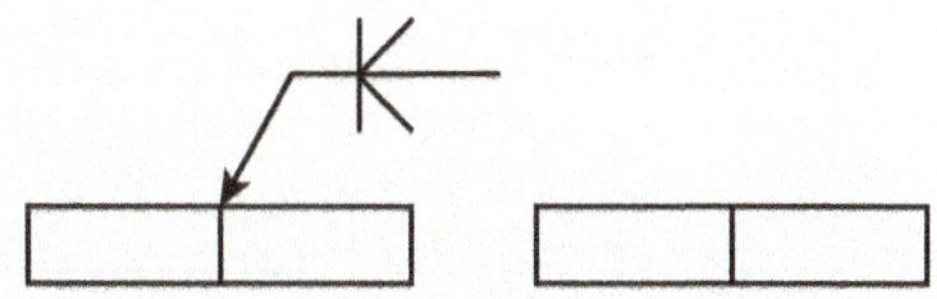

11.

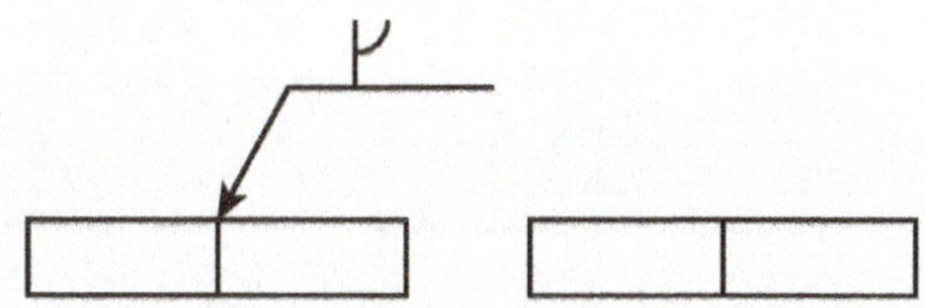

12.

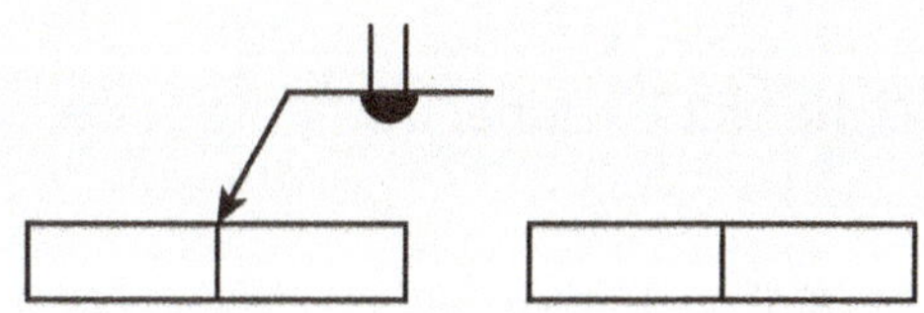

13.

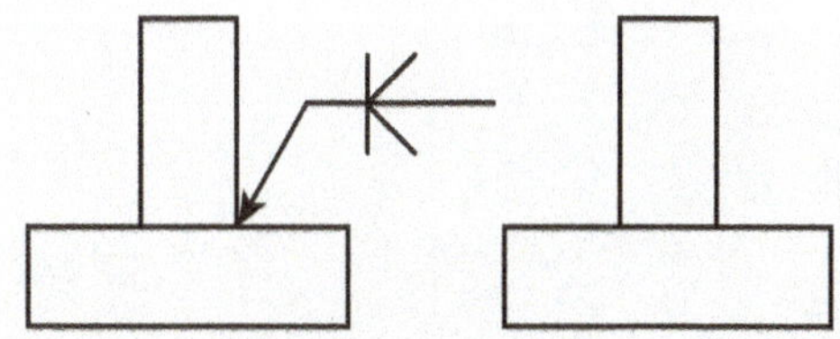

14.

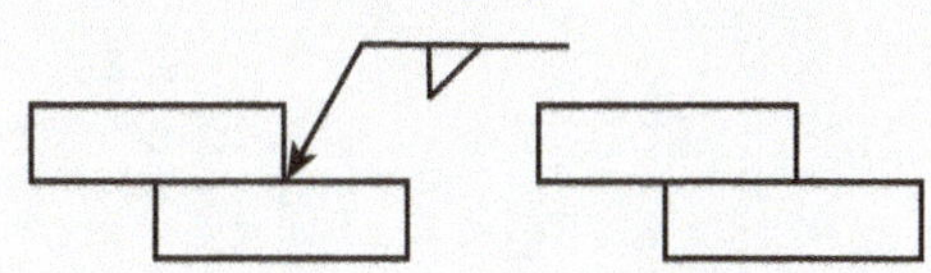

15.

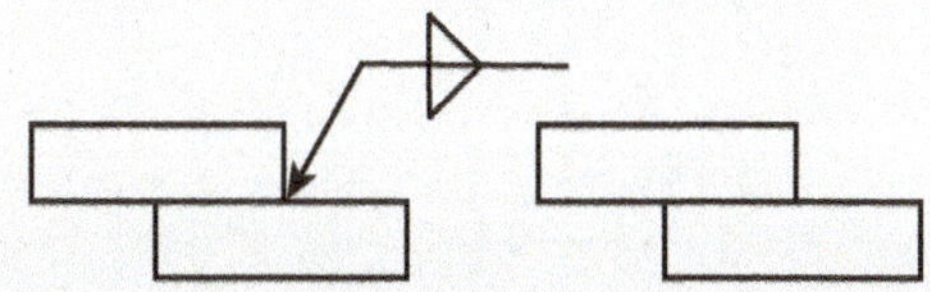

16.

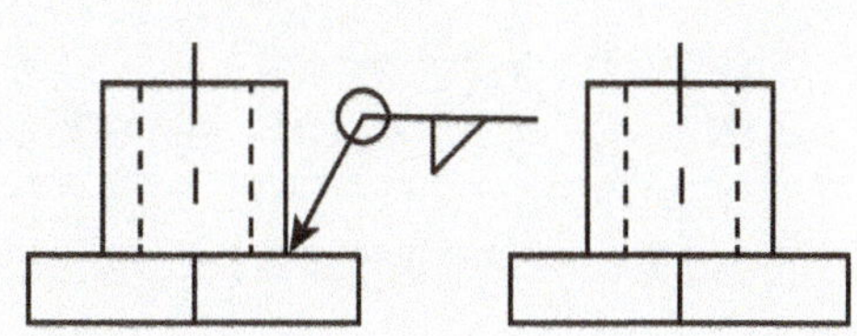

Review Exercises

Draw the symbol that represents the weld shown.

1. (Example)

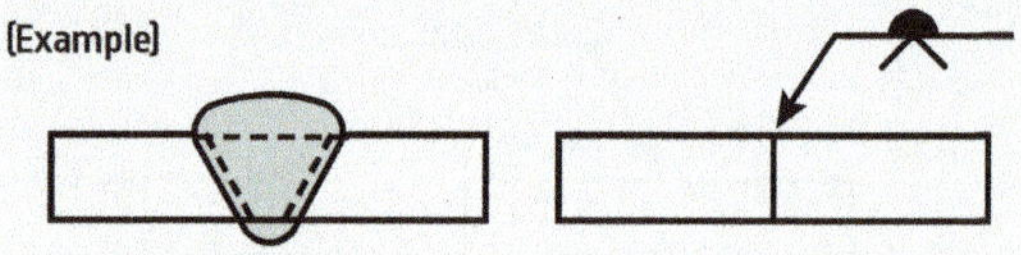

2.

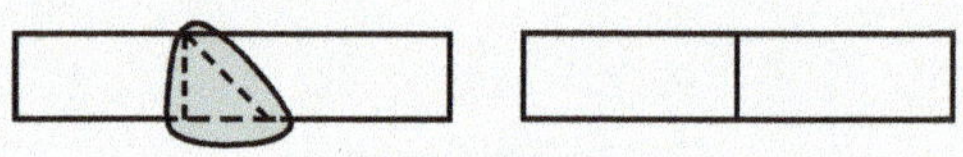

3.

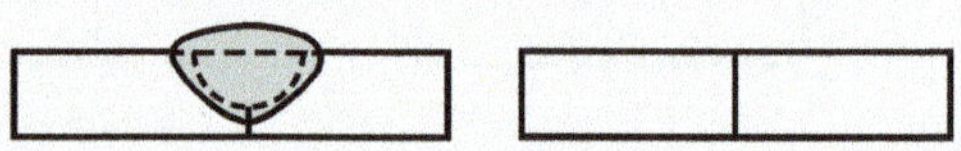

4.

5.

6.

7.

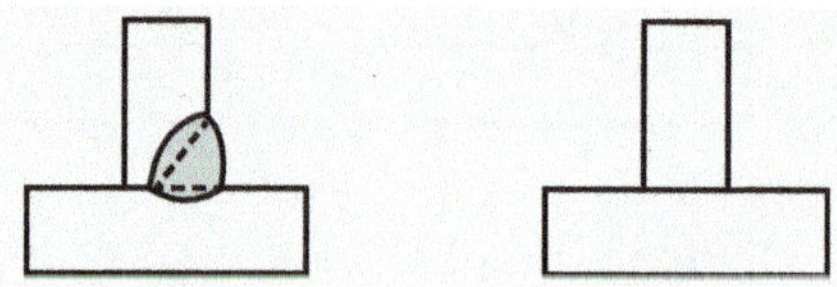

8.

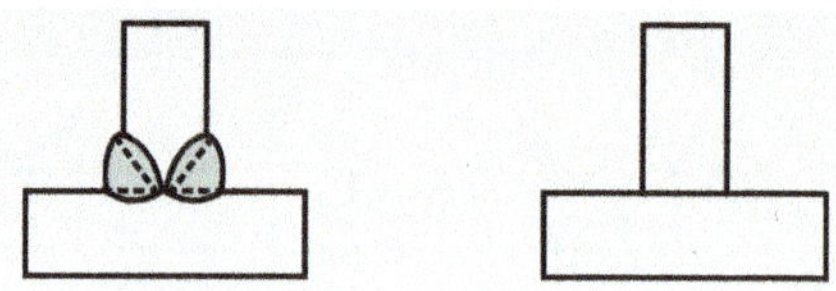

Describe the weld symbol using the correct terminology.

1. (Example)

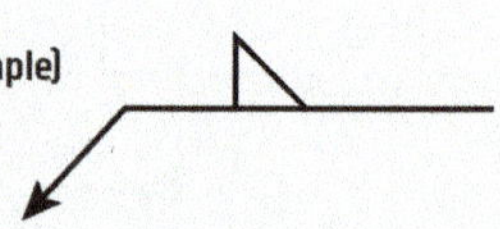

Fillet Weld

Other Side

2.

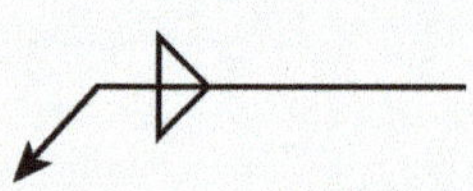

3.

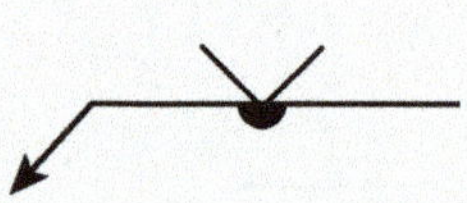

4.

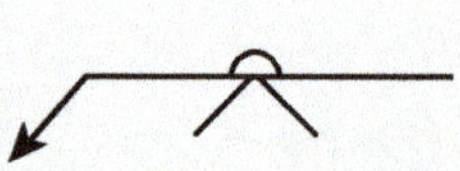

5.

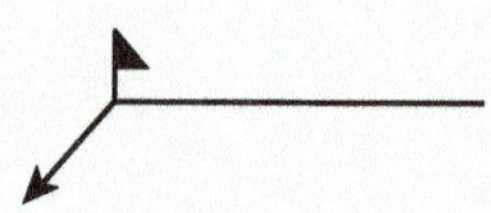

6.

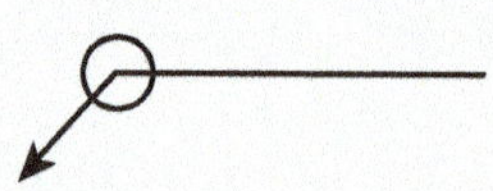

7.

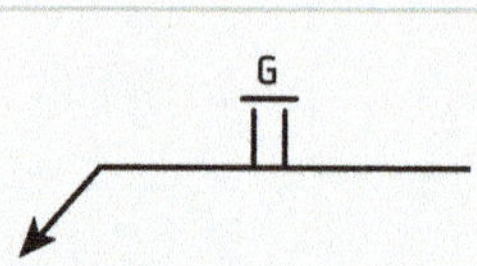

8.

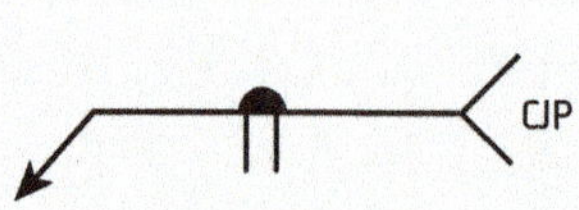

9.

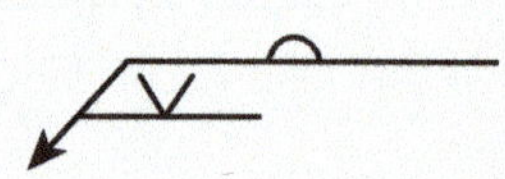

10.

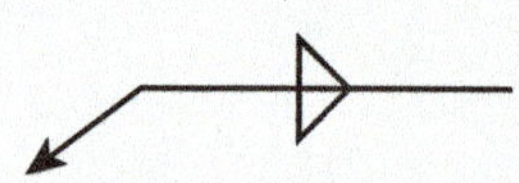

11.

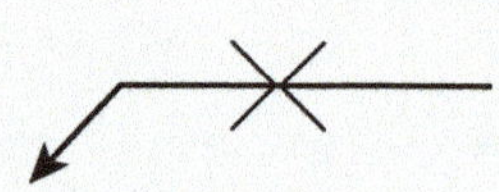

12.

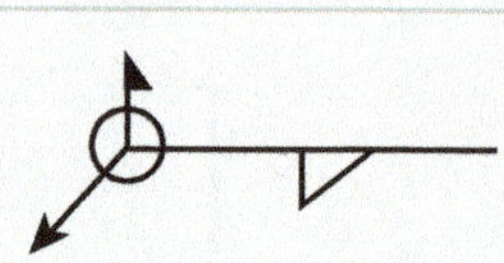

Notes

Notes

CHAPTER 3: Measuring and Tools

Chapter Objective

After learning the information in this chapter you will be able to:

- Understand the different measuring systems
- Understand fractional, decimal and metric measuring systems
- Interpret different types of measurements on drawings
- Convert measurements between the different systems
- Convert measurements between different units
- Identify the different measuring tools
- Accurately use the different measuring tools

Key Terms

Fractions
Decimals
Metric
Rulers
Dial calipers
Protractors
Converting measurements
Dual dimensioning

Welding prints are often covered in measurements, and each measurement is critical to the assembly of a part or product. Accurate measuring is an essential skill for welders. After all, we can lay down the best weld in the world, but if the part is not in the right place, we just created frustration, re-work and additional cost. Finding employees who can accurately use measuring tools and convert measurements is becoming more and more of an issue for manufacturers.

I've taught students from a wide variety of backgrounds and experiences, and my experience has generally been that about 30% of my class does not know how to use common measuring tools such as a micrometer, dial calipers or even a simple tape measure. Learning to properly read and use measuring devices will probably increase your speed and will definitely increase your accuracy. If you identify a measurement as "the small line next to the bigger line," then you may want to spend some extra time on this chapter!

Three different measuring systems are commonly used on prints. As welders, we are most likely to use either the U.S. customary measuring system or the International System of Units (SI). The US system uses fractional measurements, and SI uses millimeters (mm). As welders, we usually work within tolerances of around 1/16th of an inch. Welding prints will sometimes include decimal measurements, but decimal measurements are used more for machinists than for welders because machinists work to much tighter tolerances than welders usually do.

MEASURING TOOLS					
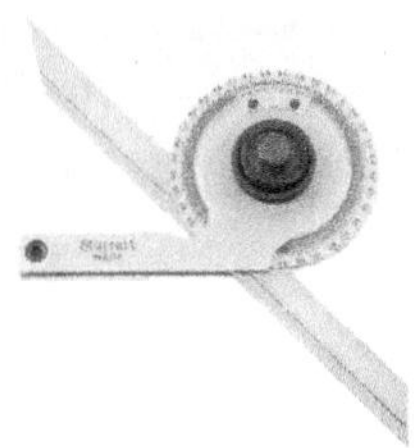	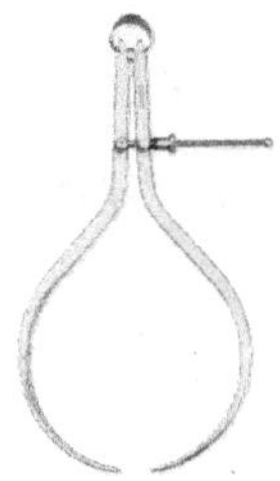	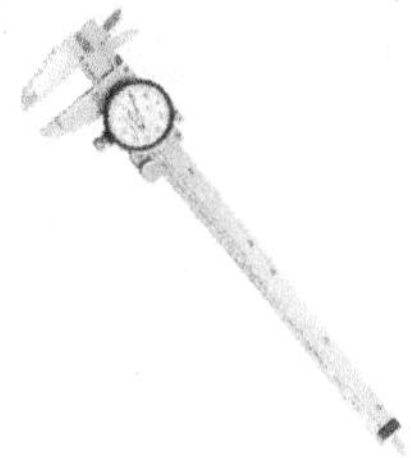			
Bevel Protractor	Caliper	Dial Caliper	Micrometer	Steel Square	Steel Tape Measure

Figure 3.1 Welders need to be able to identify and use a variety of different measuring devices. All images provided by Starrett, all rights reserved.

Regardless of the type of measurement system your shop uses, it is important to know or at least to become familiar with both the U.S. system and SI. If the drawings with which you must work are done in SI measurements, then you should acquire tools configured in millimeters. Welding prints produced using decimal measurements will often have those measurements converted to either U.S. or SI for you. Table 3.1 will be a useful tool for you whether you are working with decimals, fractions or millimeters.

INCHES TO MILLIMETERS									
in.	mm	in.	mm	in.	mm	in.	mm	in.	mm
1	25.4	21	533.4	41	1041.4	61	1549.4	81	2057.4
2	50.8	22	558.8	42	1066.8	62	1574.8	82	2082.8
3	76.2	23	584.2	43	1092.2	63	1600.2	83	2108.8
4	101.6	24	609.6	44	1117.6	64	1625.6	84	2133.6
5	127.0	25	635.0	45	1143.0	65	1651.0	85	2159.0
6	152.4	26	660.4	46	1168.4	66	1676.4	86	2184.4
7	177.8	27	685.8	47	1193.8	67	1701.8	87	2209.8
8	203.2	28	711.2	48	1219.2	68	1727.2	88	2235.2
9	228.6	29	736.6	49	1244.6	69	1752.6	89	2260.6
10	254.0	30	762.0	50	1270.0	70	1778.0	90	2286.0
11	279.4	31	787.4	51	1295.4	71	1803.4	91	2311.4
12	304.8	32	812.8	52	1320.8	72	1828.8	92	2336.8
13	330.2	33	838.2	53	1346.2	73	1854.2	93	2362.2
14	355.6	34	863.6	54	1371.6	74	1879.6	94	2387.6
15	381.0	35	889.0	55	1397.0	75	1905.0	95	2413.0
16	406.4	36	914.4	56	1422.4	76	1930.4	96	2438.4
17	431.8	37	939.8	57	1447.8	77	1955.8	97	2463.8
18	457.2	38	965.2	58	1473.2	78	1981.2	98	2489.2
19	482.6	39	990.6	59	1498.6	79	2006.6	99	2514.6
20	508.0	40	1016.0	60	1524.0	80	2032.0	100	2540.0

Table 3.1

Measuring with Fractions

In this book, we will focus on showing you the tools you will use to measure and how you can become proficient in using them. You can review adding, subtracting, multiplying and dividing fractions in a math class. I am certainly not a math instructor, but I do have some ideas that might help you better understand how to read a tape measure. Just remember, there are tools out there that work the same way but have different types of measurements.

When we learn how to use a tape measure, we only need to understand one inch at a time. Believe it or not, this is where most mistakes are made. If a part is 47 3/16 in., it is easy to read the 47 in., but that 3/16 in. measurement is often a little harder to discern. Smaller measurements are harder to determine, and as I said before, we welders usually need to be within 1/16 in. tolerance.

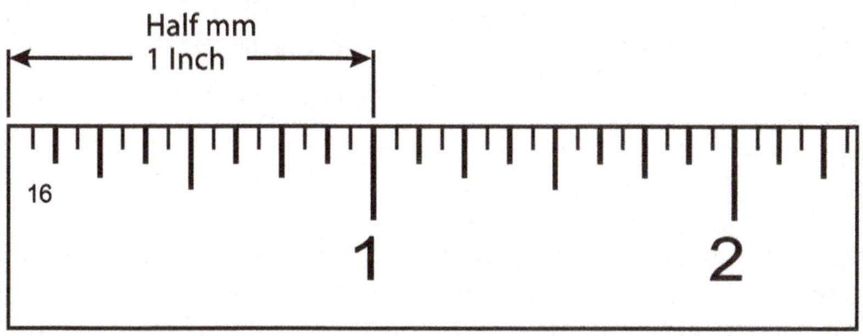

Figure 3.2

Anyone who is familiar with fractions understands that the quotient of any number divided by itself is one. If we understand that measuring is no more complicated then subdividing one inch, we can divide that inch into very simple measurements. Let's begin with the idea that the smallest denomination on most tape measures is 1/16 in. We know that there are sixteen 16ths in one inch. That means that two halves, four quarters, eight eighths, sixteen 16ths and sixty-four 64ths all equal one inch. Fractions are always reduced to lowest terms. For example, if you have a measurement of 4/8, you would reduce it to 1/2.

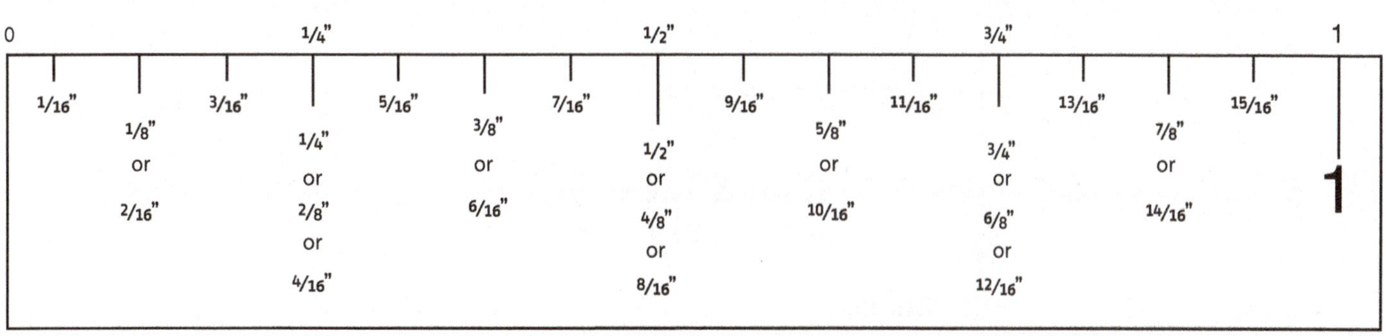

Figure 3.3

If we examine a tape measure, we will see lines of several different lengths (Fig. 3.2 and 3.3). The lines that indicate divisions of one inch are the longest, the 1/2 in. lines are next longest, then the quarter inch lines, then the 1/8 in. lines, all the way down to the shortest, which are usually the 1/16 in. lines.

Measuring with Decimals

Machinists work with the decimal system because it is much more accurate. Except for a few very special applications, welders don't generally work with measurements small enough to be measured in decimals. The decimal is very different from the fractional system but still based on one inch. With the decimal system, one inch is divided into 10 equal lengths referred to as base measurements or base units. Measuring with the decimal system is often done to one thousandth of an inch. 1/1000 in. is written in decimal form as 0.001.

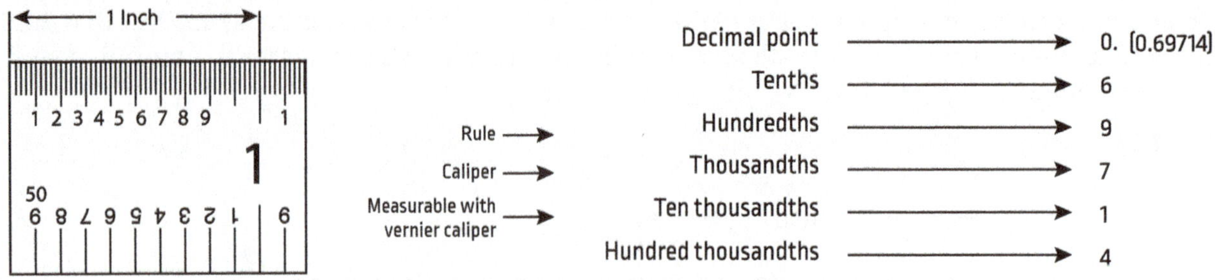

Figure 3.4 With the decimal system each line represents .02

I like to use an easy-to-remember way of subdividing the inch into smaller numbers. For example, the decimal equivalent of 1/2 in. is 0.500 in. If we divide that number in half, each half would equal 1/4 in. or 0.250 in. If we divide that by two, we end up with 1/8 in. or 0.125 in. Finally, we can divide 1/8 in. in half to get 1/16 in. or 0.0625 in. Remember, this is the smallest unit of measure you will usually encounter as a welder. In most cases, it would be perfectly acceptable to round 0.0625 in. up to 0.063 in., as most welders will not use measurements that run into the ten thousandths.

Measuring with Metrics

Measuring with millimeters is done all over the world. I have worked with SI drawings in the past, and it's my opinion that once you learn SI, you will like it. One millimeter is 1/1000 of a meter. On the majority of tape measures, measurements are marked at every centimeter (cm) (1 cm = 10 mm). One inch is equal to 25.4 mm. Some rulers or tape measures may show half millimeters, but you will not regularly use that measurement as a welder.

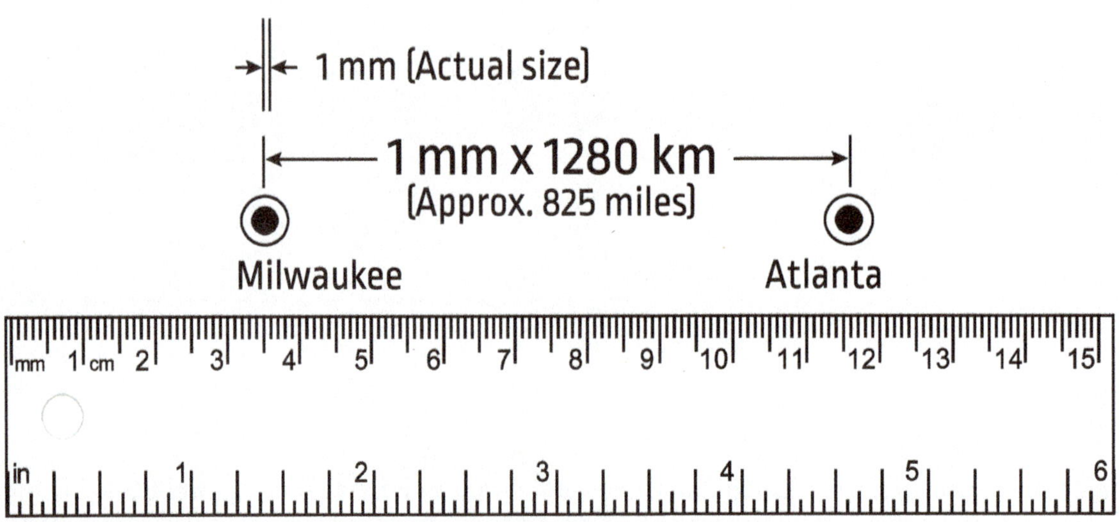

Figure 3.5 Most metric rulers are graduated in millimeters (1 thousandth of a meter). The numbers on the ruler are placed every centimeter (centimeter = 10 millimeters).

When using SI, we use prefixes with a base unit. As we move to the left or to the right of the base unit, we add prefixes (Table 3.2 and Figure 3.6).

PREFIXES, EXPONENTS, AND SYMBOLS				
Decimal Form	Exponent of Power	Prefix	Symbol	Meaning
1 000 000 000 000 000 000	=1018	Exa	E	Quintillion
1 000 000 000 000 000	=1015	Peta	P	Quadrillion
1 000 000 000 000	=1012	Tera	T	Trillion
1 000 000 000	=109	Giga	G	Billion
1 000 000	=106	Mega	M	Million
1 000	=103	Kilo	K	Thousand
100	=102	Hecto	H	Hundred
10	=101	Deca	Da	Ten
1				Base Unit
0.1	=10-1	Deci	D	Tenth
0.01	=10-2	Centi	C	Hundredth
0.001	=10-3	Milli	M	Thousandth
0.000 001	=10-6	Micro	µ	Millionth
0.000 000 001	=10-9	Nano	N	Billionth
0.000 000 000 001	=10-12	Pico	P	Trillionth
0.000 000 000 000 001	=10-15	Femto	F	Quadrillionth
0.000 000 000 000 000 001	=10-18	Atto	A	Quintillionth

Table 3.2

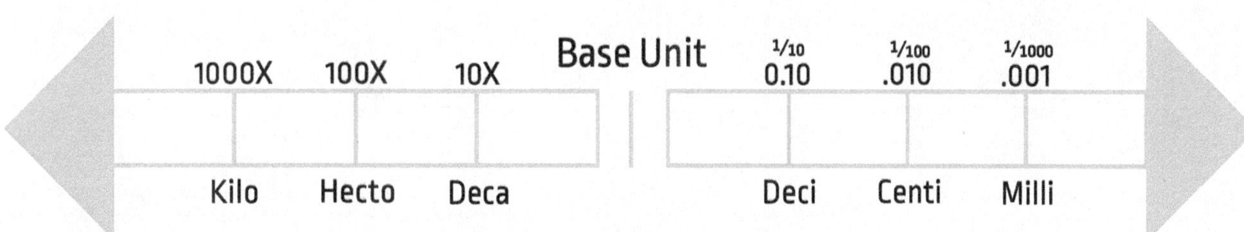

Figure 3.6 The decimal system revolves around a base unit.

Welders sometimes need to convert measurements from millimeters to fractions or decimal measurements and vice versa. There are many formulas that we can use to get the job done. Most fabrication shops that sell their products outside of the US will also need to work with SI.

Protractors

Protractors are used to measure angles. There are many different types of protractors out there. Metal protractors are best for welders because they're the least likely to melt or get burned. Angles are measured in degrees on all protractors. Protractors usually have two scales that measure 0°-180° or 0°-90°, but some have just one 0°-180° scale.

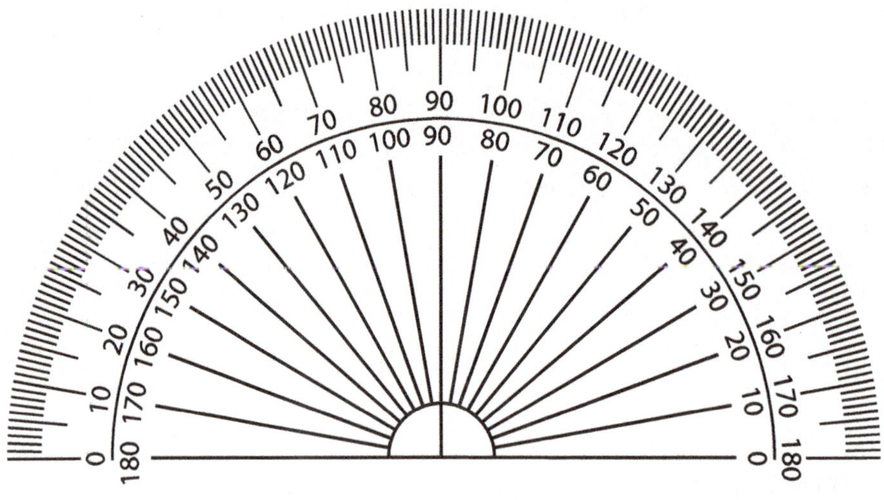

Figure 3.7 Protractors are usually based on a 180° scale or broken into two 90° scales.

All protractors have a starting point, or a point of origin from which the measurement starts. The point of origin may vary depending on the type of protractor being used. You must make sure the point of origin is aligned with the angle being measured.

Figure 3.8 Moveable arm protractor.

Converting Measurements

Converting Inches to Feet - We divide the number of inches by 12.

Example: **168 in. ÷ 12 = 14 ft.**

Converting Feet to Inches - We multiply the number of feet by 12.

Example: **20 ft. × 12 = 240 in.**

Converting Inches to Millimeters - We multiply the number of inches by 25.4 (25.4 mm. = one in.).

Example: **22 in. × 25.4 = 558.8 mm.**

Converting Millimeters to Inches - We divide the number of millimeters by 25.4.

Example: **365 mm. ÷ 25.4 = 14.40 in.**

Converting Fractions to Decimals - We divide the top number by the lower number and usually round the number to three decimal places.

Example: **1/2 = 2 ÷ 4= .500**

Converting Decimals to Fractions - Converting a decimal to a fraction is a little bit more involved because fractions and decimals are not exactly the same measurement. Again, we welders very rarely work with measurements this small. If you find the need to do this conversion, refer to the conversion chart provided.

Review Exercises

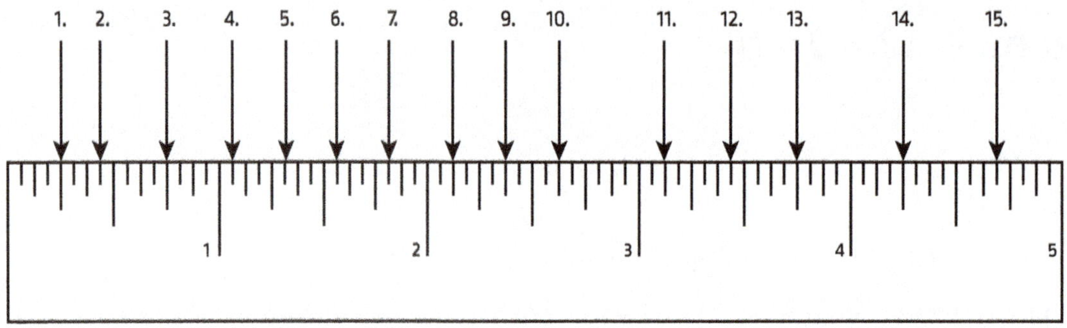

Provide the fractional measurement on the answer lines provided.

1. ____________ 4. ____________ 7. ____________ 10. ____________ 13. ____________

2. ____________ 5. ____________ 8. ____________ 11. ____________ 14. ____________

3. ____________ 6. ____________ 9. ____________ 12. ____________ 15. ____________

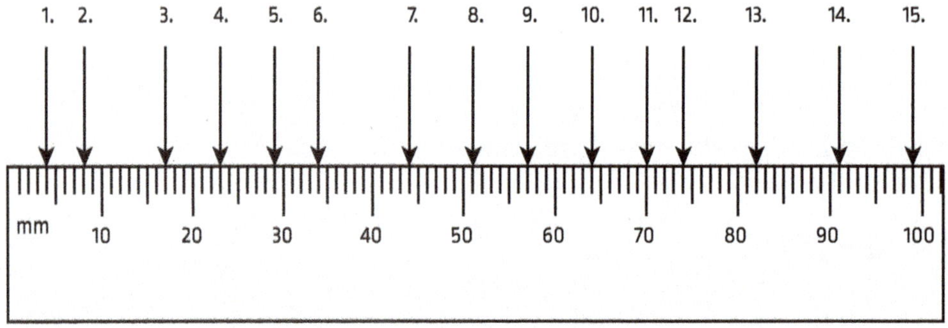

Provide the metric measurement on the answer lines provided.

1. ____________ 4. ____________ 7. ____________ 10. ____________ 13. ____________

2. ____________ 5. ____________ 8. ____________ 11. ____________ 14. ____________

3. ____________ 6. ____________ 9. ____________ 12. ____________ 15. ____________

Review Exercises

Determine the fractional measurement with a 1/16 in. graduated ruler. Once finished, convert the fractional measurement to metric measurements using the formula: 1 in. x 25.4 = mm. Round to the nearest millimeter.

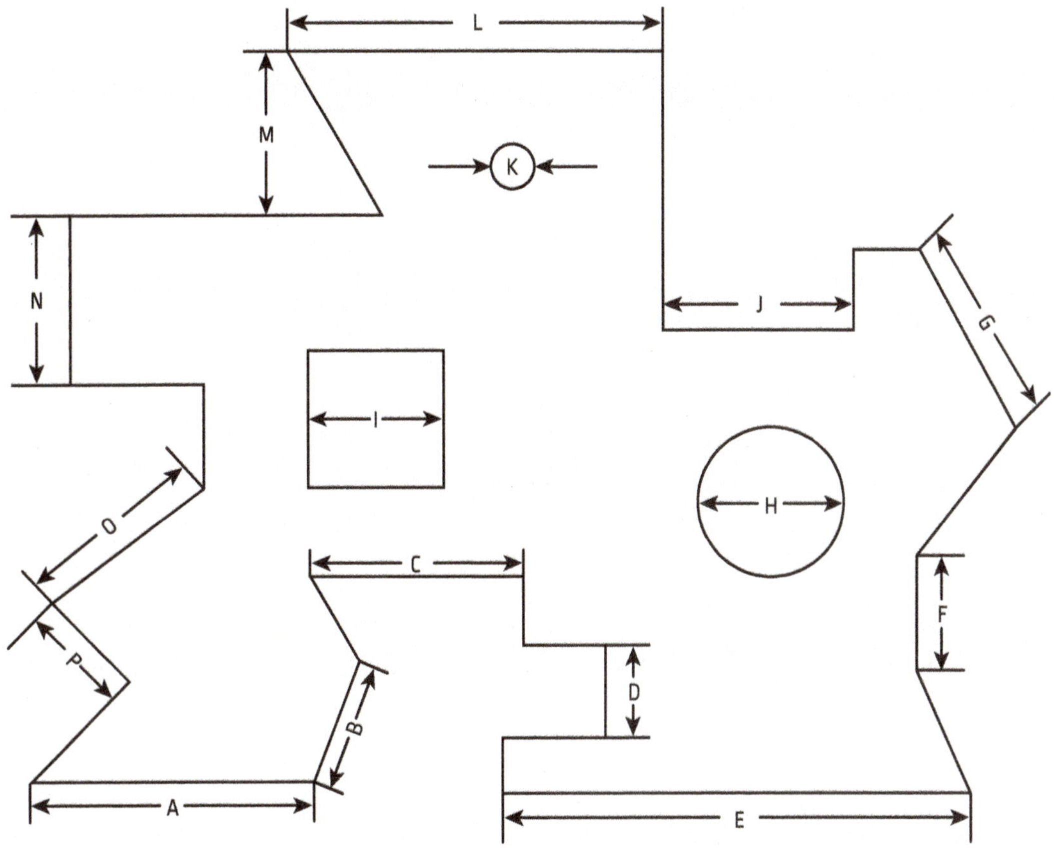

Provide the fractional measurement.

A. ______
B. ______
C. ______
D. ______
E. ______
F. ______
G. ______
H. ______
I. ______
J. ______
K. ______
L. ______
M. ______
N. ______
O. ______
P. ______

Convert fractional measurement to metric measurement.

Notes

CHAPTER 4: Welding Processes

Chapter Objective

After learning the information in this chapter you will be able to:

- Identify the designation for the major welding processes
- Understand the difference between the major welding processes
- Describe the different equipment used for each process
- Compare the advantages and disadvantages of each welding process
- Understand the designation abbreviations for each welding process

Key Terms

Soldering (S)
Brazing (B)
Oxyfuel (OFW)
Gas Metal Arc Welding (GMAW)
Gas Tungsten Arc Welding (GTAW)
Shielded Metal Arc Welding (SMAW)
Flux Cored Arc Welding (FCAW)
Submerged Arc Welding (SAW)
Stud Welding (SW)
Resistance Welding (RW)
Resistance Spot Welding (RSEW)
Projection Welding (PW)

Most of us are familiar with the four major welding processes commonly known as Gas Metal Arc Welding (GMAW), Gas Tungsten Arc Welding (GTAW), Shielded Metal Arc Welding (SMAW), and Flux-Cored Arc Welding (FCAW). In this chapter we will introduce you to more welding processes and their proper industry names. Each process has an AWS letter designation that we will include to help you understand industry language. The welding processes you use will depend on the type of shop you work in and the products you build. You will often use a combination of welding processes to create a part or product.

This chapter will also cover soldering and brazing. Soldering and brazing are not considered welding because they are performed at a lower temperature than welding. In addition, the base metal is not melted in soldering or brazing, and the filler material melts at a lower temperature than the base material. Materials that are soldered or brazed differ in strength from welded materials, but this is exactly why we use them.

Soldering (S)

Soldering is a joining method that uses non-ferrous filler that melts at a much lower temperature than the base metal. It is often referred to as soft soldering. Soldering is used to join the copper tubing in the plumbing in your house. It is used to connect wiring and occasionally thin sheet metal. Soldering is used for low strength, low temperature (below 840° F/449° C) and minimal pressure applications.

Figure 4.1a Soldering.

Figure 4.1b Brazing.

Brazing (B)

The biggest difference between brazing and soldering is that brazing is performed at temperatures above 840° F/449° C, whereas soldering is performed below these temperatures. Like soldering, brazing does not melt the base metal; however, brazing produces a stronger finished product than soldering. You can generally rely on brazing to produce a strong joint, depending on the alloy of the filler material and the joint type. For this reason, brazing is commonly used to repair broken cast iron or steel parts. I learned how to fix cast iron using this process years ago, and I've gone on to perform many repairs of this type with great success. Cast iron tends to be brittle, but the relatively low heat level of brazing keeps the cast iron from becoming too brittle after welding. Note that you can usually skip preheating and post heating when brazing cast iron.

Brazing relies on a "capillary action"; that is, the filler material is actually sucked down into the weld joint. Braze welding, a related but different process, uses special joint types to ensure the capillary action needed for a solid joint. Brazing is generally automated, but braze welding is performed by manually feeding the filler material into the joint.

Oxyfuel Gas Welding

This process uses a mixture of a fuel gas and oxygen to produce a flame capable of melting metal and creating strong welds on most ferrous and nonferrous materials. The most common fuel gas used for gas welding is acetylene. For metals with lower melting points, such as aluminum and lead, hydrogen is typically used. Gases such as propane or MAPP™ gas (methylacetylene-propadiene propane) can also be used, but they are more commonly used for brazing. Oxyfuel gas welding equipment can also be used to cut materials and to heat them for bending or forming. You are most likely to encounter oxyfuel gas welding in the automotive industry, where it is used to weld thin sheet materials. This process is also popular for welding sections of thin tubing, like those used in the inner structure of airplanes.

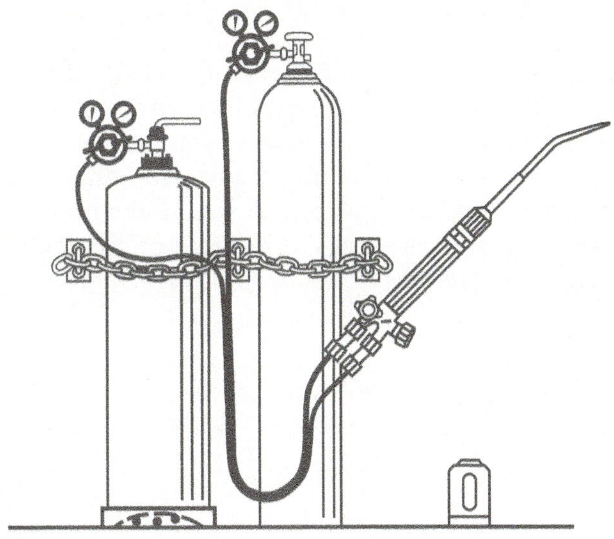

Figure 4.2 An Oxyfuel setup.

Gas Metal Arc Welding (GMAW)

Gas Metal Arc Welding is widely known as MIG welding, but the correct AWS designation is GMAW. GMAW is probably the most widely used welding process in manufacturing today. It is easy to learn, versatile, fast and can produce excellent welds in several different types of ferrous and non-ferrous materials such as steel, stainless steel and aluminum. GMAW is considered a semiautomatic welding process because the welder moves the welding gun across the weld joint by hand while the filler metal is fed automatically from a large spool. It is easy to make GMAW fully automated using robotics.

GMAW uses an electric arc to melt the base metal (Fig. 4.3). The filler metal is fed into the puddle automatically by the wire feeding mechanism on the welding power source. GMAW uses different methods of metal transfer including short circuit, globular, spray and pulsed.

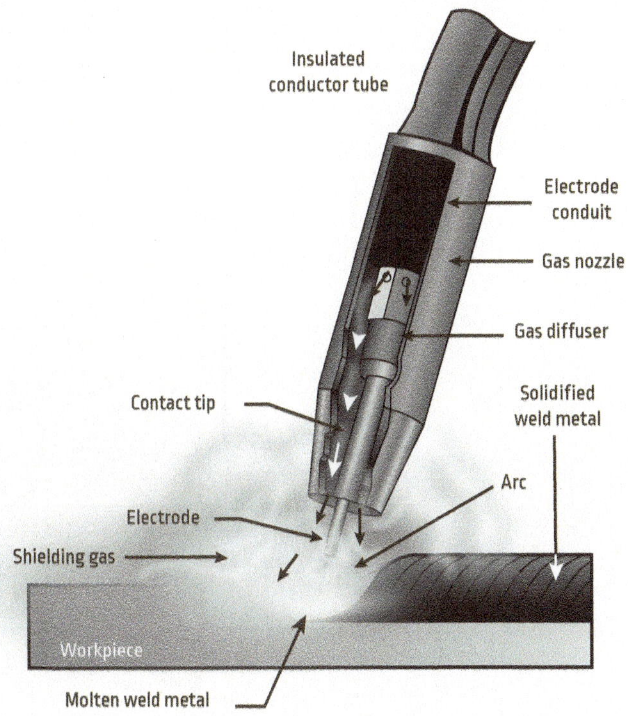

Figure 4.3 The GMAW process.

Flux Cored Arc Welding (FCAW)

FCAW uses the same machinery as GMAW. The major difference between the two processes is the wire. Flux cored wire is a hollow sheath filled with flux on the inside. The flux protects the weld and helps hold the weld in place during out of position welding. Flux can also contain alloying elements needed to create a particular type of weld. The FCAW process has two different types of wire classifications. One is gas-shielded, and the other is self-shielded wire. Gas-shielded wire uses both flux and a cover gas to protect the weld (Fig. 4.4). Shelf shielded wire uses no gas; it has flux on the inside of the wire, allowing it to be used outdoors without fear of losing cover gas.

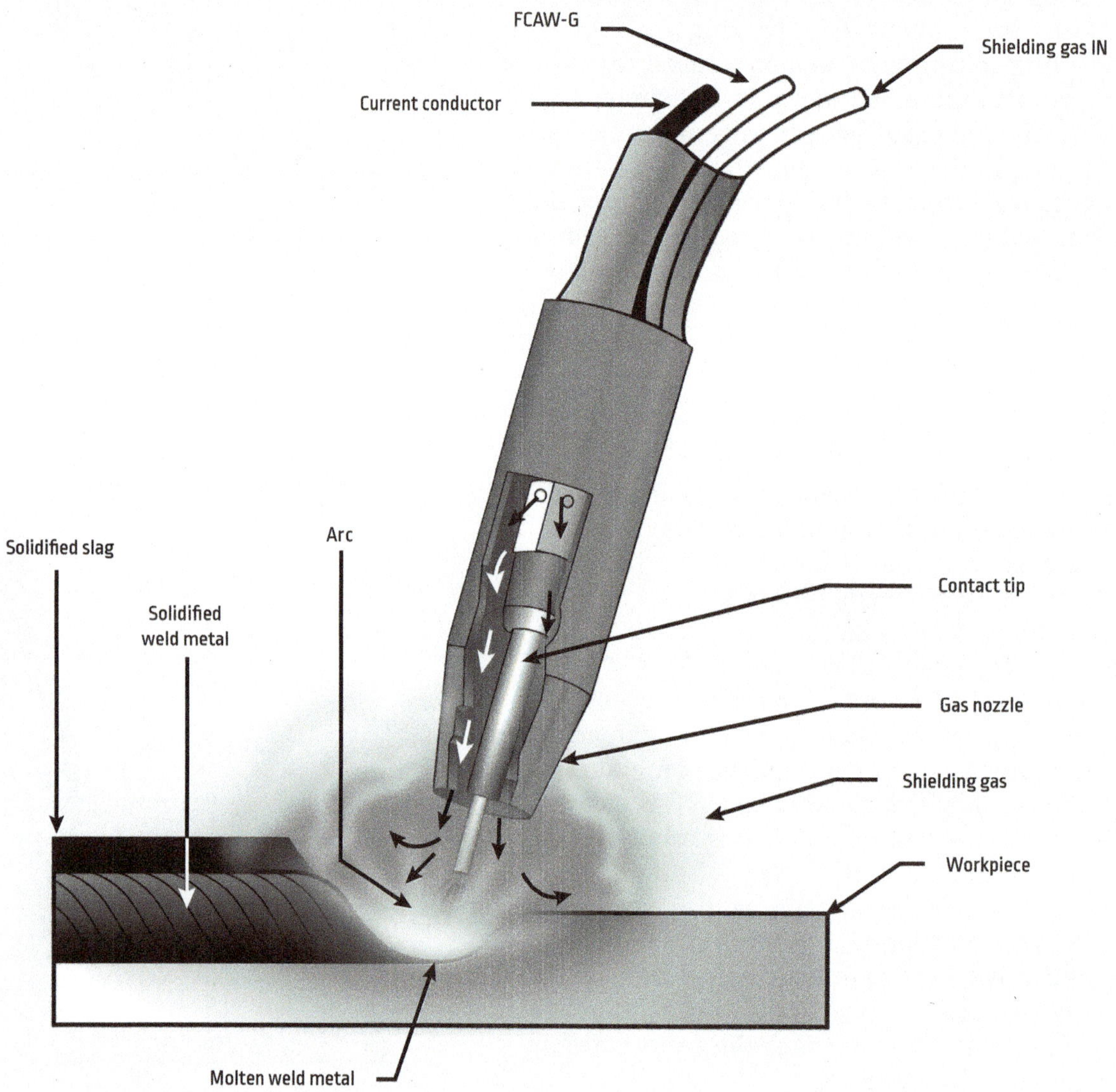

Figure 4.4 The FCAW-G process.

FCAW creates a very fluid puddle with deep penetration characteristics. It has a high deposition rate and is generally used on thicker sections of base metal. It operates at higher voltage and amperage ranges than GMAW. FCAW can be semiautomatic or fully automated, just like GMAW.

Shielded Metal Arc Welding (SMAW)

SMAW is a process that uses a filler rod covered in different types of flux. An electric arc melts both the base metal and the filler metal at the same time, creating a good weld with excellent strength characteristics. SMAW has a relatively low operating factor because the slag (melted flux) has to be chipped and cleaned after each stick of welding rod is burned. Even though it's slower than GMAW and FCAW, SMAW has desirable qualities like its low cost and the fact that it's easy to learn. SMAW is popular and is still widely used in the construction industry and for maintenance repair welding. Most welding performed outdoors is SMAW.

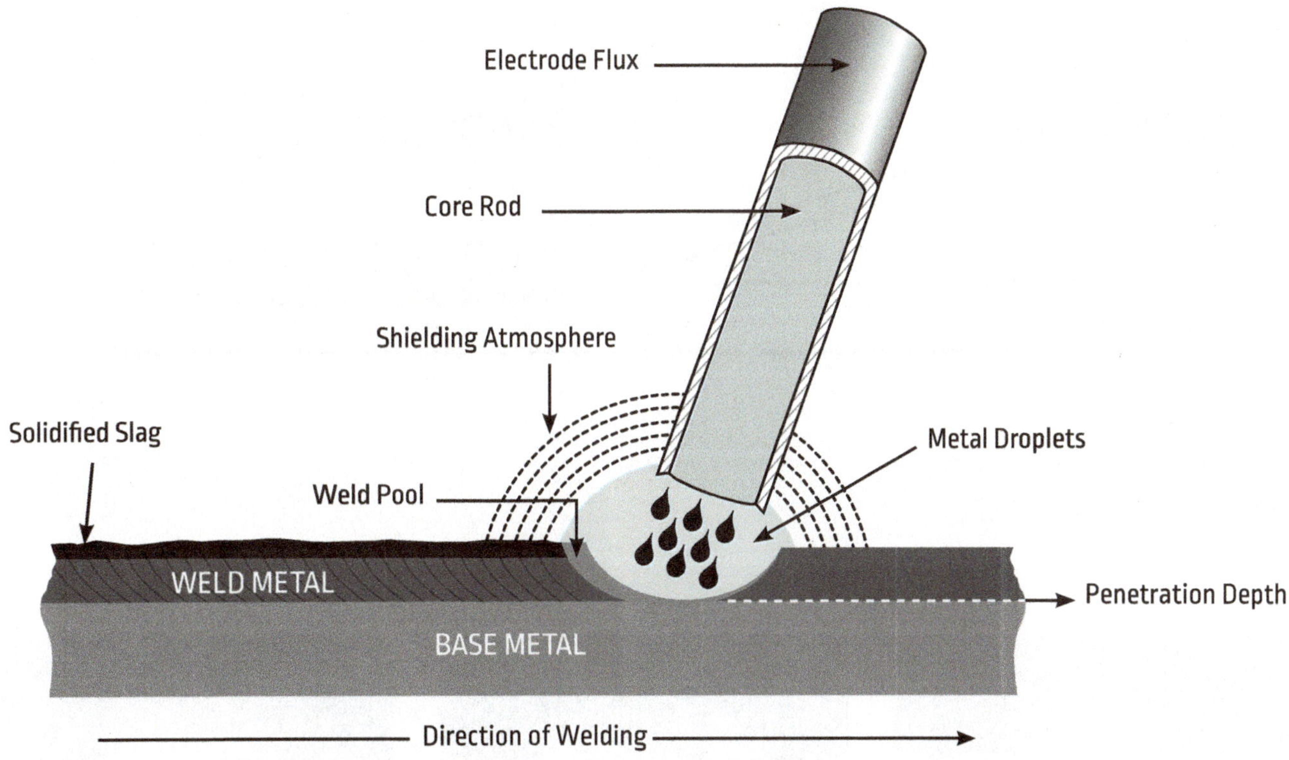

Figure 4.5 The SMAW process.

Gas Tungsten Arc Welding (GTAW)

GTAW is commonly referred to as TIG. It was initially known as Heliarc because the process was developed with helium as the cover gas. Argon eventually replaced helium because of its lower cost and smoother arc. The electrode is made from tungsten alloy, which has a high melting point. GTAW has a low deposition rate, therefore, it's considered slow in comparison to the other processes. Though GTAW is slow, it has many other desirable qualities. GTAW is a clean process that can produce high quality welds. It can be used to weld a variety of different metals merely by changing the filler material or polarity. Figure 4.6 shows the GTAW process and GTAW welds on several different base metals.

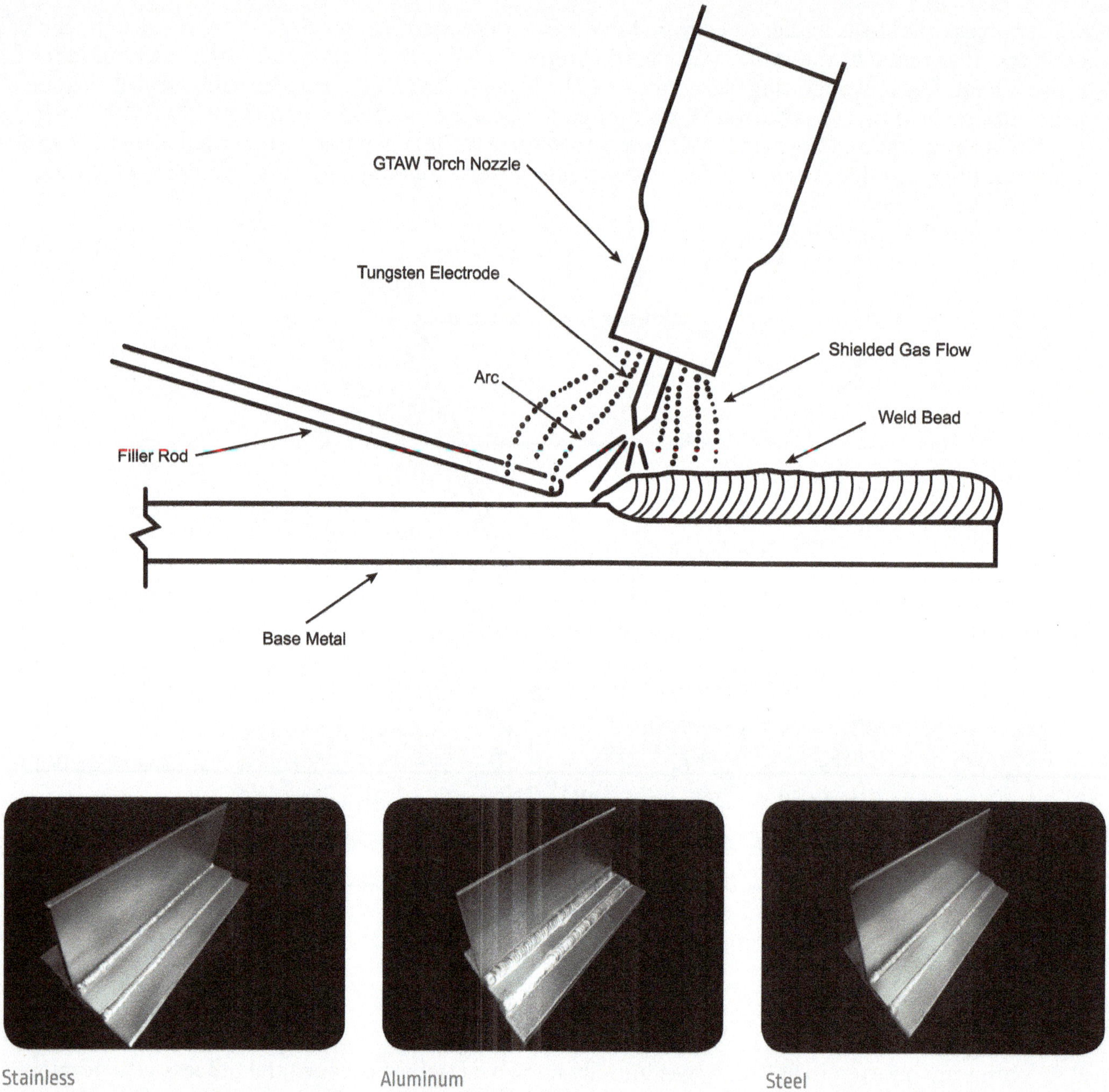

Figure 4.6 The GTAW process and different types of welds on stainless, aluminum and steel.

Submerged Arc Welding (SAW)

"Submerged arc" brings to mind some sort of underwater welding, but that is not an accurate depiction of this process. This process is known as submerged arc welding because the weld pool is completely blanketed under a layer of granular flux that pours out of a hopper positioned above the weld joint (Fig. 4.7). The flux shields the weld pool from the atmosphere. Because the weld is hidden under the flux, SAW produces very little smoke, fumes or spatter. SAW is used to weld thick sections of materials, or to weld thinner sections while moving at a fast travel speed. SAW can be hand held or automated. Because of the granular flux, SAW can only be performed in the flat and horizontal positions. SAW has a high deposition rate and runs at relatively high amperage and voltage ranges.

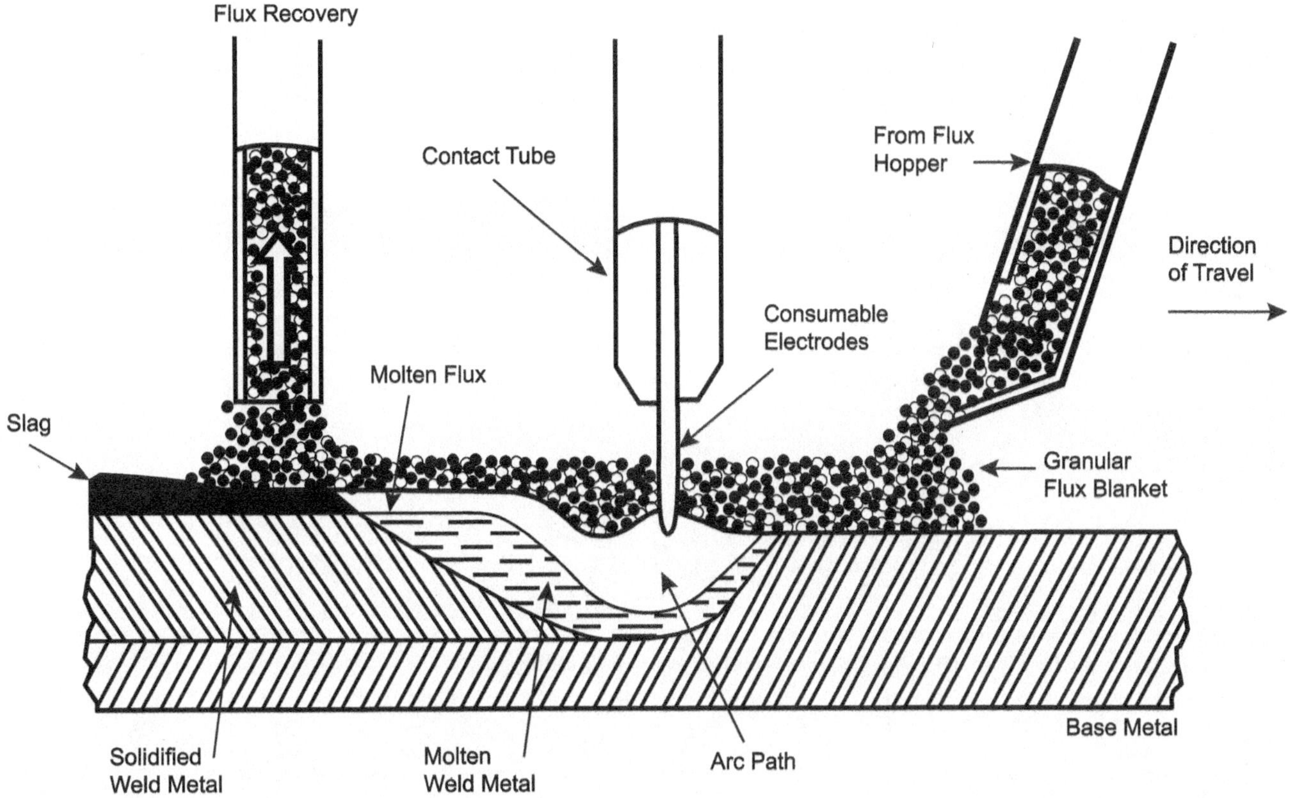

Figure 4.7 The SAW process.

Figure 4.8 Hand held SAW.

Stud Welding (SW)

Stud welding is a process where the arc and the melting of the base metal and the stud happen at the same time. Stud welds are primarily used to attach a threaded bolt to flat steel as a fastener. Stud welds are often seen on bridges for construction purposes like holding the concrete to the bridge or bolting forms together for concrete forming.

Figure 4.9a Stud welding setup. Provided by Nelson Stud Welding, Inc., all rights reserved.

Figure 4.9b Stud weld. Provided by Nelson Stud Welding, Inc., all rights reserved.

Resistance Welding (RW)

Resistance welds are another group of welds done for different applications. The process is usually performed without filler metal. Instead it uses the resistance of the material to generate enough heat to melt metal together.

Resistance Spot Welds (RSW)

Resistance spot welds are made with copper electrode arms that come together over two pieces of material and apply pressure on the parts. Current is then passed through the electrode, causing resistance between the metal and forming a weld nugget with no filler needed. RSW machines can be hand held or highly automated and are predominantly used in the automotive industry. Most of the welds on your car were probably formed using resistance spot welding. Figure 4.10 shows an RSW setup.

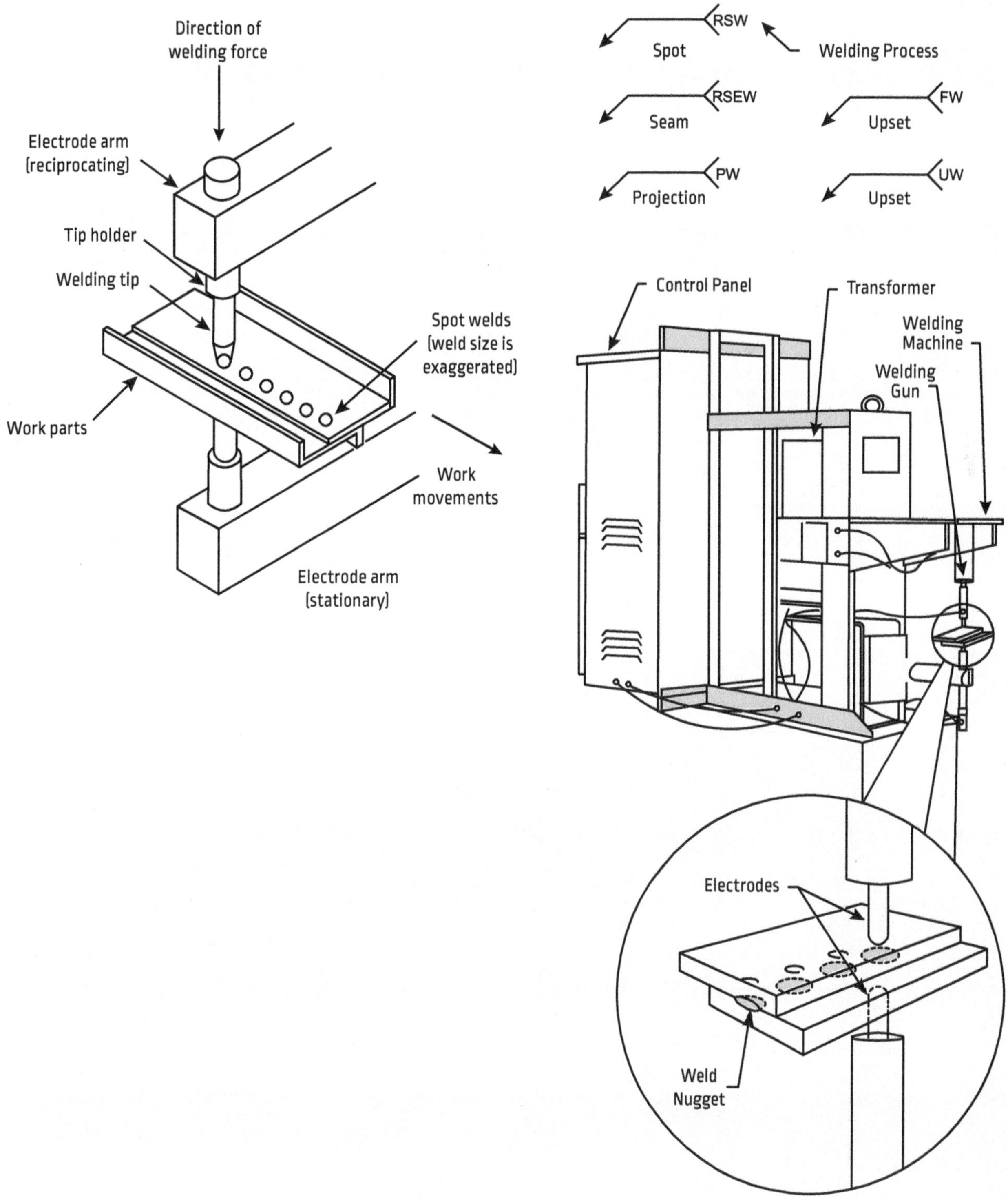

Figure 4.10 Resistance spot welding setup.

Resistance Seam Welds (RSEW)

RSEW is similar to spot welding because it conducts current through electrodes and creates resistance between the parts. The electrodes work the same as spot welding but are round in shape (Fig. 4.11) and create a resistance weld that is flat and smooth. Resistance seam welds are used in many different applications. The fuel tank in your car was probably resistance seam welded around the edges.

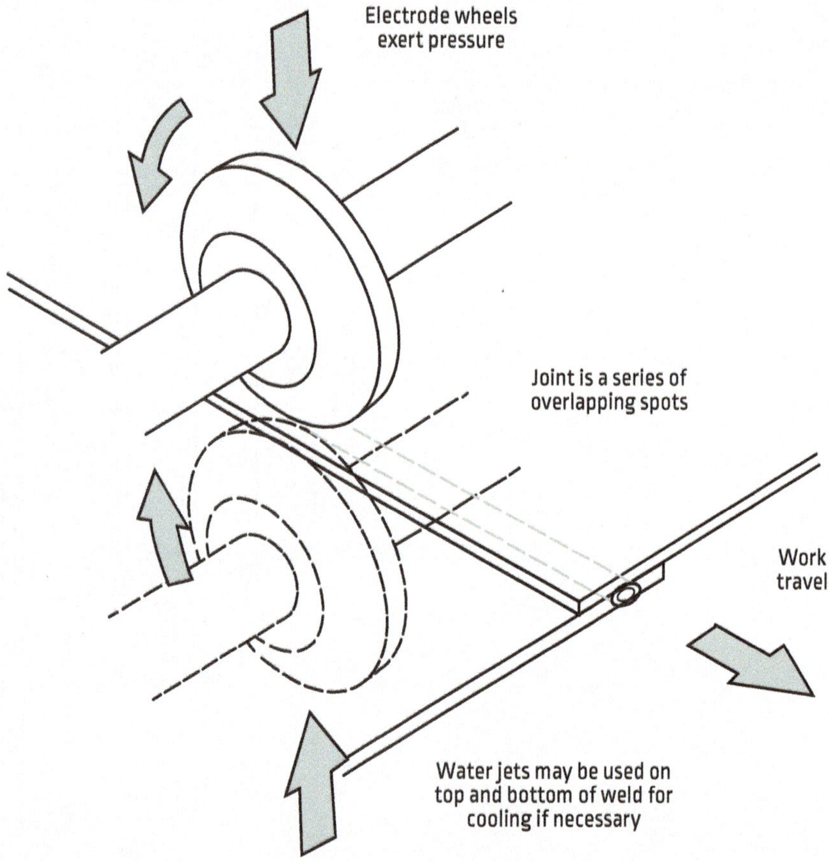

Figure 4.11 Resistance Seam Welding (RSEW) is a process in which a long series of overlapping spot welds create one continuous weld.

Projection Welds (PW)

Projection welds work the same way as a spot weld. The big difference is that one part usually has a dent "projected" out in one direction. This pushes a round area outward. The result is that the pieces do not sit tightly against each other like a spot weld does. The weld nugget happens only where the surfaces of the dents are touching. The result is that the piece of material that has the "projection areas" is lifted off from the base metal and welded only in the raised areas.

Review Questions

Match the letter to the process designation.

1.	____	This process uses temperatures below 840° F	A	GMAW
2.	____	A welding process that uses a solid wire as the filler metal and the electrode	B	GTAW
3.	____	A welding process that clamps material and fuses a weld nugget using electrical resistance, time and pressure	C	SMAW
4.	____	This welding process is used in the flat position and can be done safely with no welding helmet	D	FCAW - Gas-Shielded
5.	____	This welding process uses a non-consumable electrode	E	GTAW
6.	____	This joining uses an Oxyfuel torch setup and relies on "capillary action" to join materials	F	Submerged Arc Welding
7.	____	This joining process uses temperatures above 840° F but does not melt the base material	G	Soldering
8.	____	This process uses a gun and electrical resistance to weld studs and threaded fasteners to base material	H	SMAW
9.	____	This process uses a Tungsten electrode	I	Brazing
10.	____	An electrical arc process used primarily outdoors because of a flux covered electrode	J	Oxyfuel Setup
11.	____	One of the oldest welding processes that is commonly used for outdoor pipeline welding	K	GTAW
12.	____	A welding and cutting process that uses a combination of a fuel, gas and oxygen	L	Submerged Arc Welding
13.	____	A welding process that uses no filler but does use electrical resistance to create enough heat to fuse metal together	M	FCAW - Self-shielded
14.	____	A welding process widely used in the automotive industry	N	IRSW
15.	____	This electric welding process uses flux inside the wire and a cover gas	O	SMAW
16.	____	This welding process uses a flux covered electrode and has the ability to weld steel, stainless steel, castings, and even aluminum	P	Resistance Spot Welding
17.	____	This process uses rollers and resistance to create a continuous fusion weld	Q	Oxyfuel Welding
18.	____	An electric arc process that has three different metal transfer methods (short circuit, globular, spray)	R	STUD Welding
19.	____	This process uses acetylene, MAPP or propane gas to melt the base metal	S	GMAW
20.	____	A welding process that uses DCT, DC- and AC, and the filler metal is fed by hand	T	Resistance Welding
21.	____	A welding process that uses an electrode/filler wire and the arc is buried in a granular flux	U	Brazing
22.	____	This electric arc process uses a tubular wire filled with flux, and no cover gas is needed	V	Resistance Seam Welding

Notes

CHAPTER 5: Fillet Welds

Chapter Objective

After learning the information in this chapter you will be able to:

- Identify the parts of a fillet weld
- Understand how to measure fillet welds
- Compare intermittent and staggered fillet welds
- Determine weld length

Key Terms

Legs	Intermittent	Size
Toes	Staggered	Lap
Throat	Pitch	Tee
Face	Length	Chain

Fillet welds are the most common and usually the most economical type of weld we find in industry. Many companies spend a great deal of money building fixtures and buying positioners to give themselves the ability to put any part into a flat or horizontal position. This makes fillet welds much easier to perform while greatly reducing the cost.

Many of the elements of fillet welds are the same for other welds. Most fillet welds need to have equal leg measurements. If the legs of the weld are different in size, they are referred to as un-equal leg fillet welds.

The size of the fillet weld is always noted on the left of the weld symbol. Weld sizes are generally given in fractional form or decimal form, but many also use millimeters. If the desired weld size is not on the symbol, you can usually find it elsewhere on the drawing. Look for a fillet weld symbol with "TYP" written in the tail section. When you see a "TYP," that means the specified size of the weld is typical to all the other welds of that type on the drawing. In other words, if there are several welds of the same type, one is given a size and the rest are to be made the same, which is indicated with "TYP".

One important part of understanding fillet welds is developing a working knowledge of the parts of a fillet weld: the vertical leg, horizontal leg, upper toe, lower toe, weld root, actual throat and weld face (Fig. 5.1). The parts of a fillet weld are the same in a lap joint and a T-joint. Remember, both a lap joint and a T-joint are formed using a fillet weld. The difference between the two is the configuration of the weld joint.

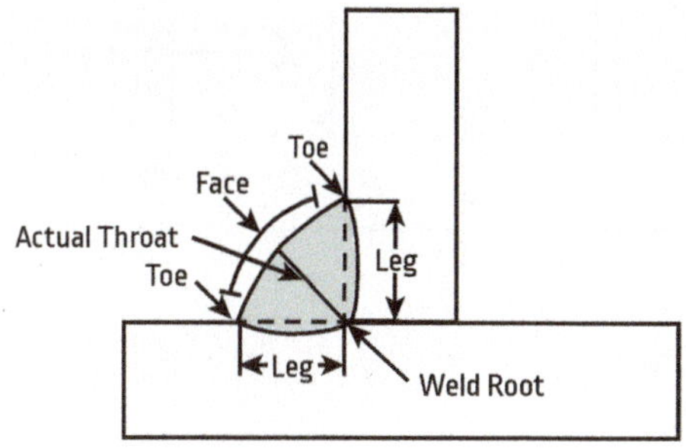

Figure 5.1 Parts of a Fillet Weld

Fillet Welds	
Print	Weld Example

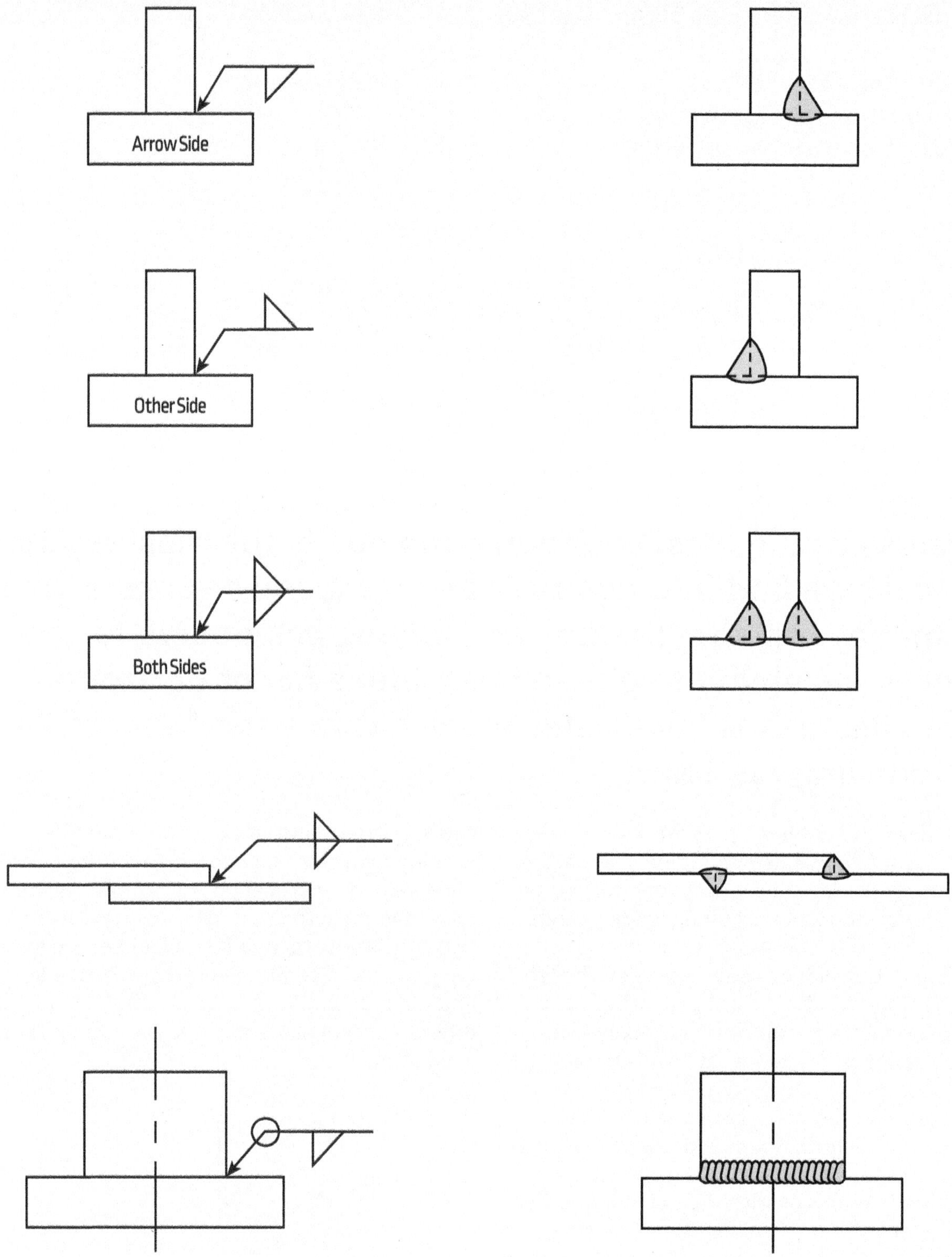

Figure 5.2 When the weld symbol is placed below the reference line, the weld is placed on the arrow side. When the weld symbol is placed above the reference line, the weld is placed on the other side of the joint. When symbols are on both sides of the reference line, the welds are placed on both sides of the joint.

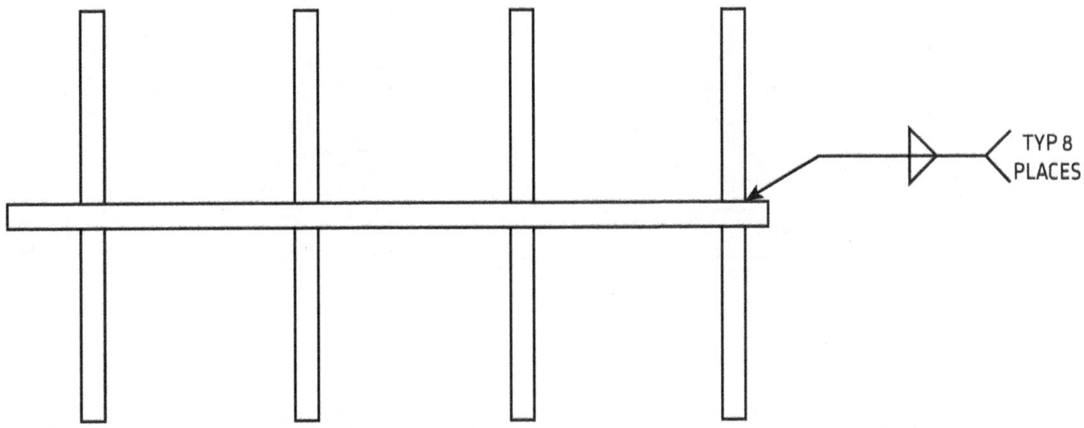

Figure 5.3 The tail of the welding symbol often contains important information. TYP in the tail section indicates that the same weld is to be placed in all four places on both sides of the joint.

Fillet Welds	
Print	Weld Example

Figure 5.4 The weld size is always placed on the left side of the symbol. The vertical line of the symbol is always on the left.

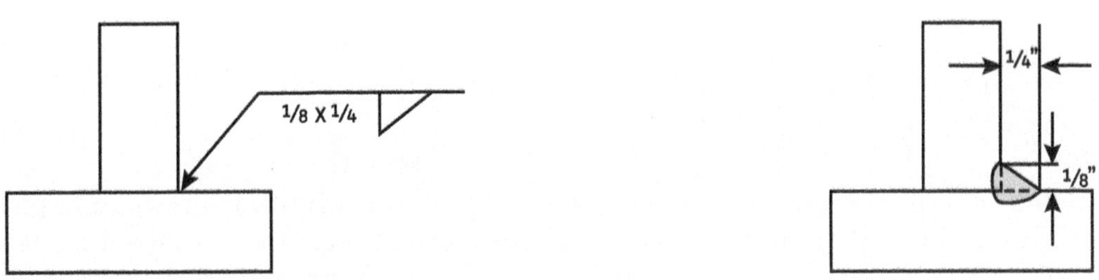

Figure 5.5 When fillet welds have unequal legs, the smaller leg is usually given first.

Figure 5.6 When fillet welds are on both sides, weld sizes are given for both welds. These welds have equal legs.

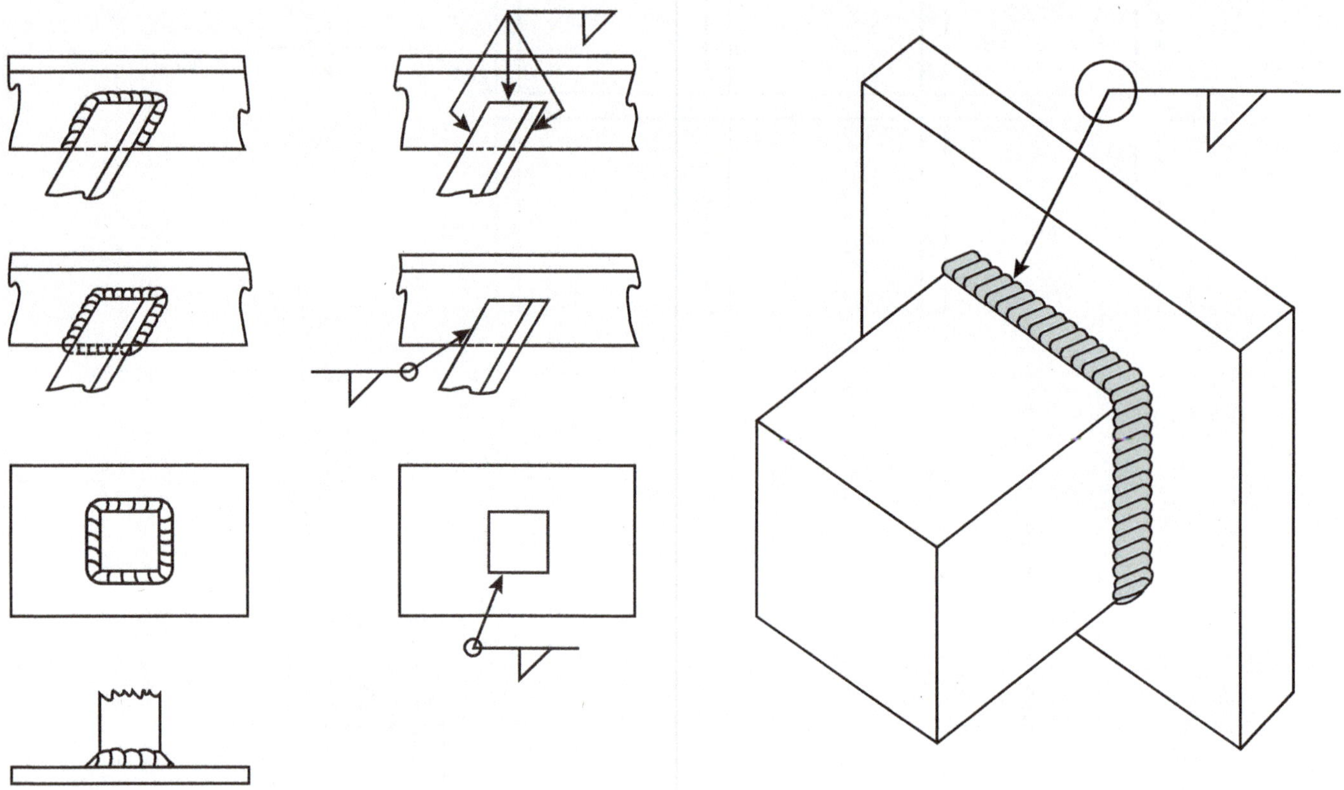

Figure 5.7 Designation of extent of welding.

Intermittent Fillet Welds

Designers and engineers spend a good deal of time determining how much weld is needed to give the part its required strength. Often, a part only needs a certain amount of weld to meet specifications. Using intermittent fillet welds, when possible, provides significant cost savings compared to welding the whole seam. Intermittent fillet welds are welds that are not solid; they are measured with intentional spacing between shorter welds. These welds often create the same amount of integrity in the product as being welded solid would.

Intermittent fillet welds are comprised of two measurements, the length and the pitch. These measurements are always on the right side of the weld symbol and are separated by a hyphen (Fig. 5.8). The first number is the length of the weld, and the second number is the center-to-center distance. A common mistake made by welders is interpreting the pitch as a measurement of the space between the welds. This is not correct; in fact, the pitch is measured from the center of one weld to the center of the next weld.

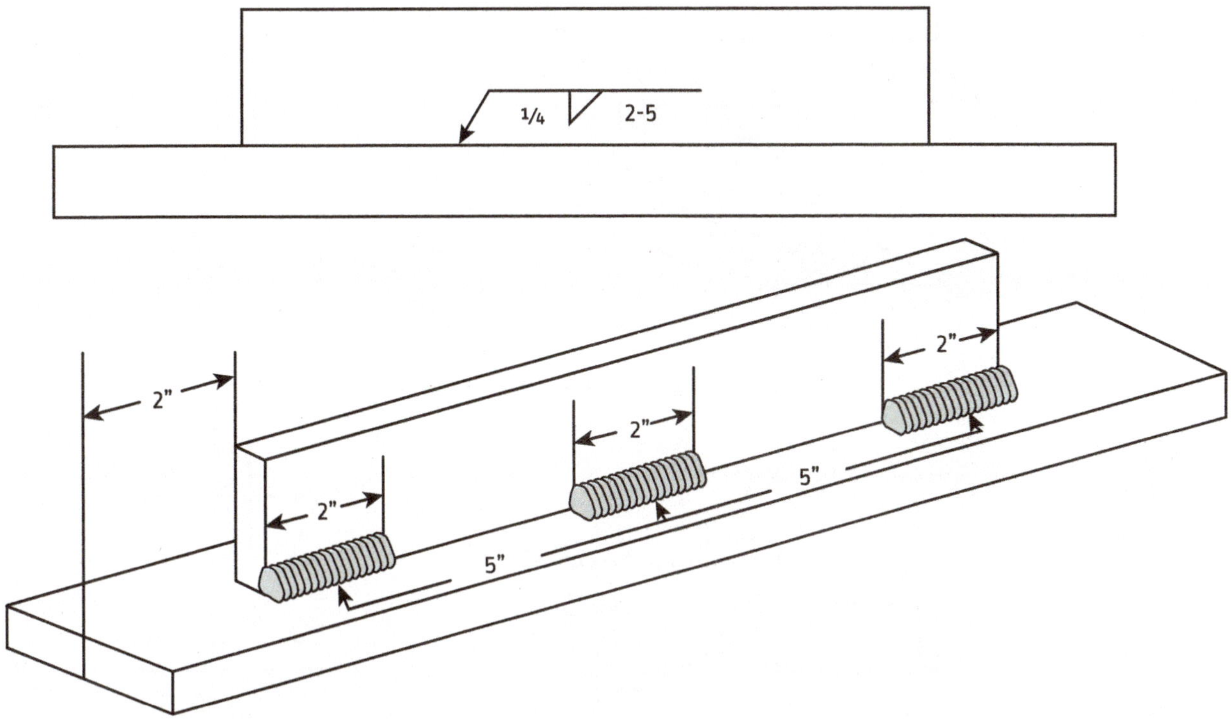

Figure 5.8 This is an intermittent fillet weld on a T-joint. The weld size is 1/4 in. These are 2 in. welds on 5 in. centers. This drawing also indicates where the first weld begins. If the start point is not indicated, the weld begins on the edge of the material.

Most intermittent fillet welds start on the edge of the weld joint, but sometimes starting at the edge may make the weld fit unevenly in the joint. In order to place the welds evenly, you might need to move in from the end to start the first weld. The first weld should generally be "moved in" the distance provided on the welding symbol.

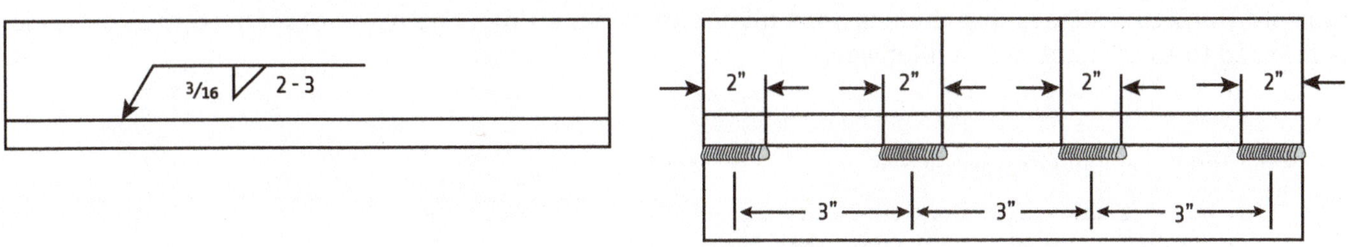

Figure 5.9 Sometimes a continuous weld is not needed. Using intermittent welds can reduce cost, distortion and repair.

Chain Intermittent Fillet Welds

Chain intermittent fillet welds are fillet welds made on both sides of the joint (Fig. 5.10). The welds are straight across from each other symmetrically on each side of the joint. Chain intermittent fillet welds reduce the total cost of the weld and help to control heat input into the base material, which results in less distortion, cost and rework.

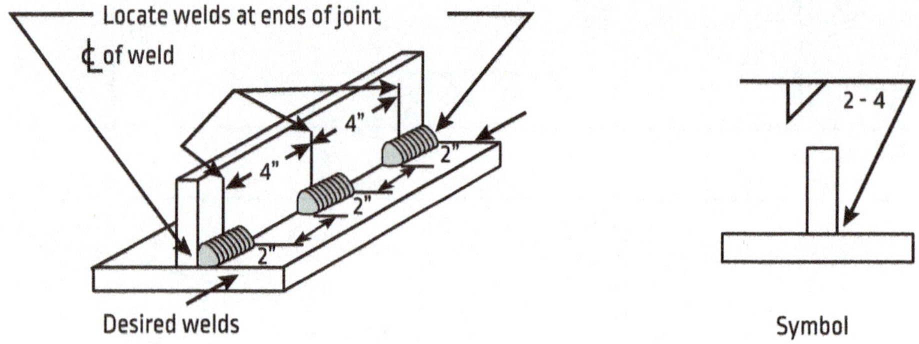

Length and pitch of increments of intermittent welding

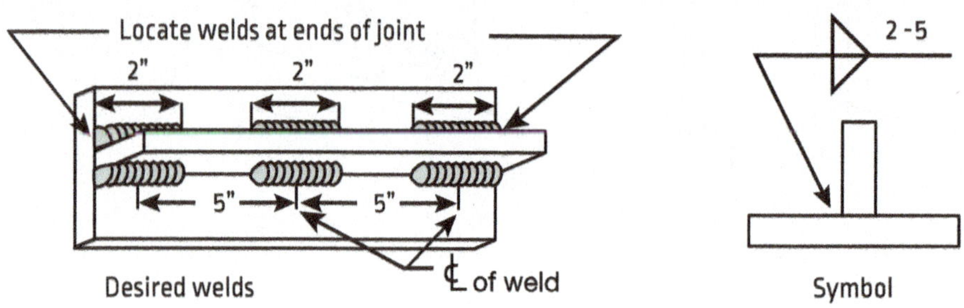

Length and pitch of increments of intermittent welding

Figure 5.10 Chain intermittent welds.

Staggered Intermittent Fillet Welds

These welds are similar to chain intermittent fillet welds. You can see this in figure 5.11. Staggered welds are formed on both sides of the joint, but they are placed in opposite positions on each side. These welds can provide the same amount of strength as welding the joint all the way across. For staggered intermittent fillet welds, the fillet weld symbol will be on both sides of the reference line but not in alignment with the symbol on the other side.

Lap joints, corner joints and T-joints can be welded using a fillet, groove weld or both types of welds. The important thing to realize is that there is not a fillet joint or lap weld. A T-joint is not referred to as a T-weld—it is referred to as a T-joint with a fillet weld.

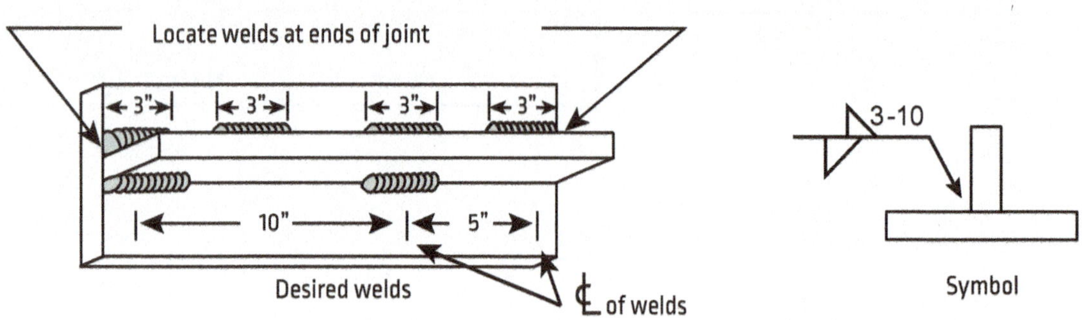

Length and pitch of increments of intermittent welding

Figure 5.11 Illustration showing staggered intermittent welds.

Joints Used with Fillet Welds

T-joints

T-joints are most commonly welded with a fillet weld and are probably one of the most widely used types of joint. A T-joint is created when one piece of material sits upright at a 90° angle to the material it is sitting on. Depending on the thickness of the material, a T-joint could require a single pass weld or a multiple pass weld. T-joints are used in all positions and in many settings including construction, structural steel and manufacturing.

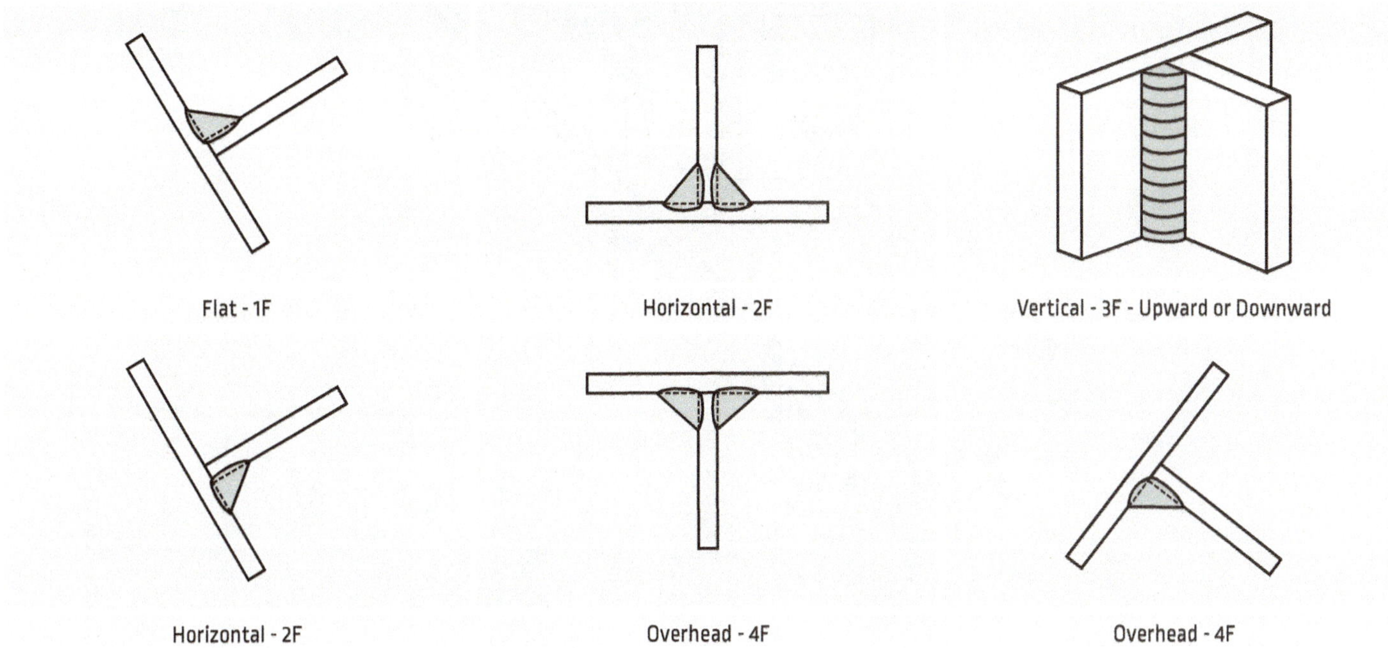

Figure 5.12 T-joints in different positions; single pass and multi pass welds.

Lap Joints

A lap joint is another common weld joint. The name lap joint describes what the weld joint looks like: two pieces of material laying on top of each other, creating an L-shaped joint. If you look at the lap joint and the T-joint closely, you'll realize they are the same joint. Just like T-joints, lap joints can be welded with single pass or multiple pass welds. Lap joints are used in all positions. Remember, "lap" is the type of joint and not the type of weld. It is a fillet weld in a lap joint. The size of the fillet weld will be noted on the left of the weld symbol.

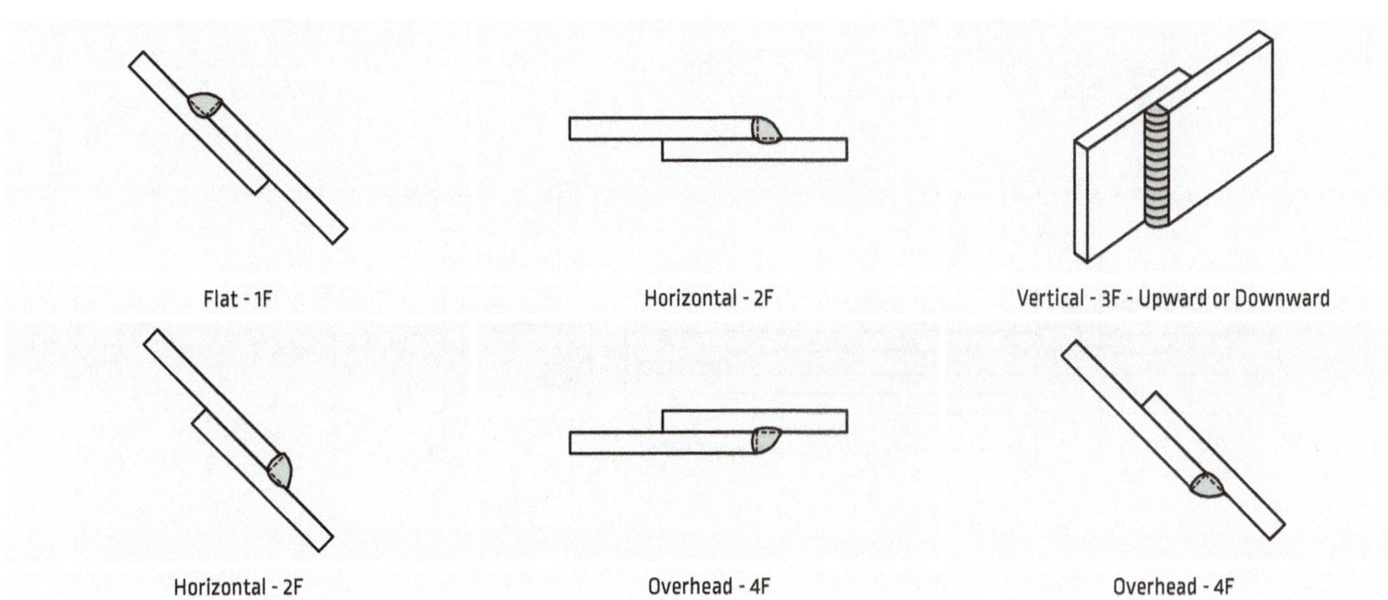

Figure 5.13 Examples of lap joints in different positions.

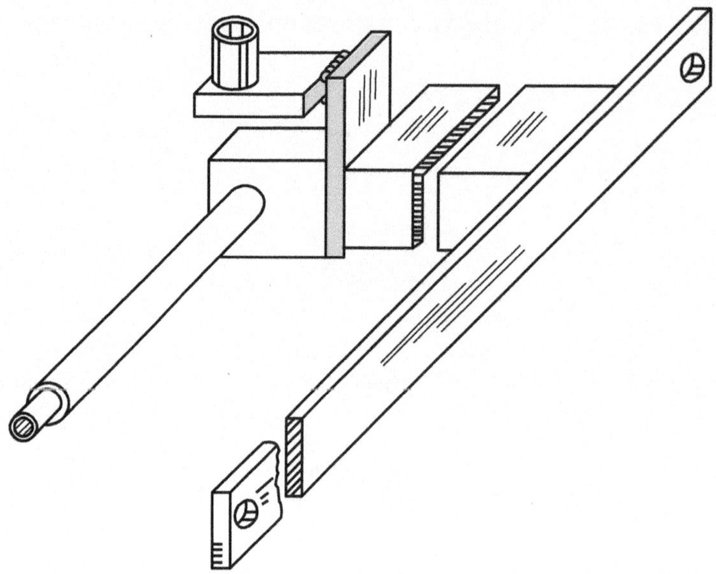

Figure 5.13 (Continued) This part makes use of lap joints.

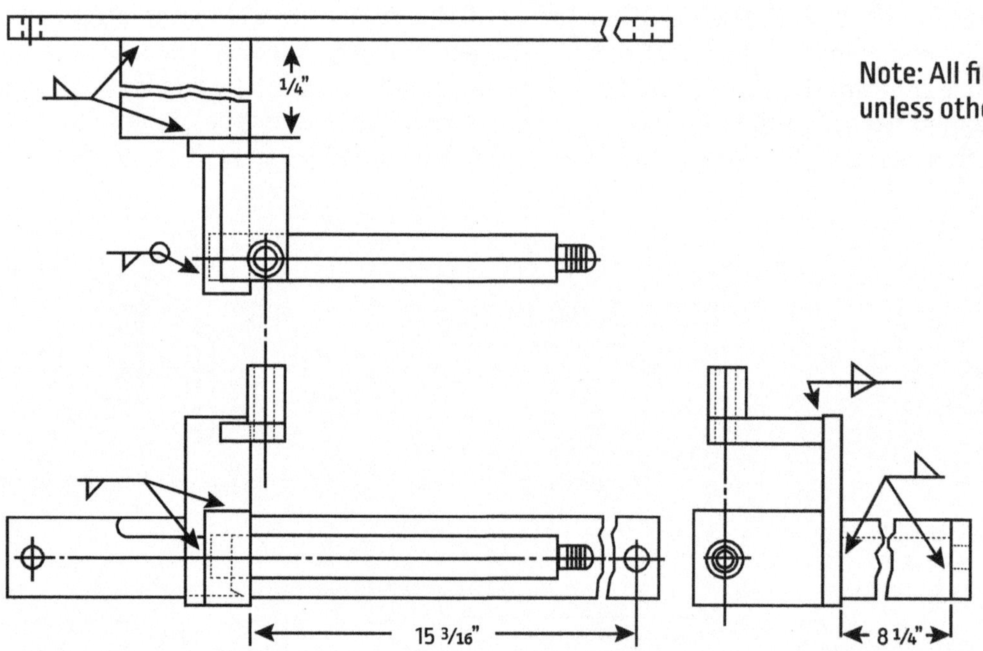

Note: All fillets 3/16"
unless otherwise specified

Figure 5.13 (Continued)

Review Questions

Draw the correct symbol to represent the weld being shown.

1. 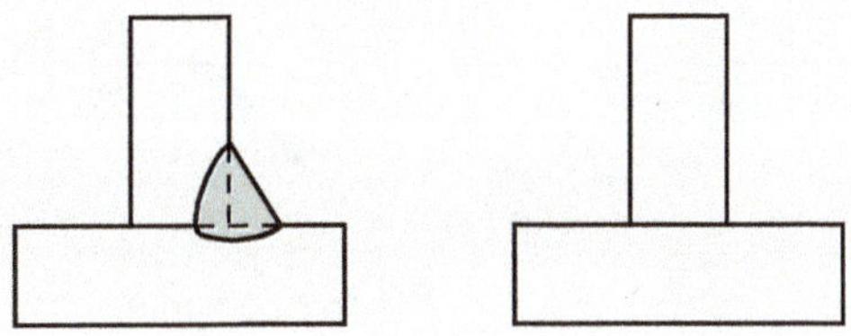

2. 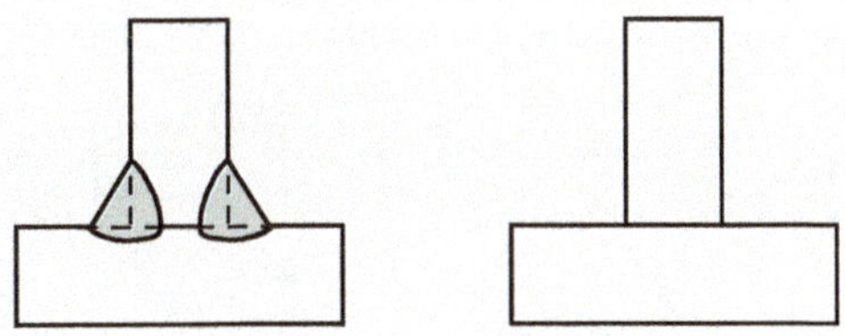

3. 

4. 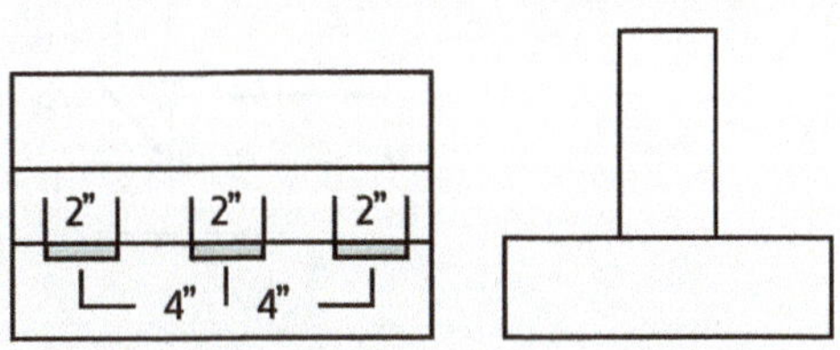

5. 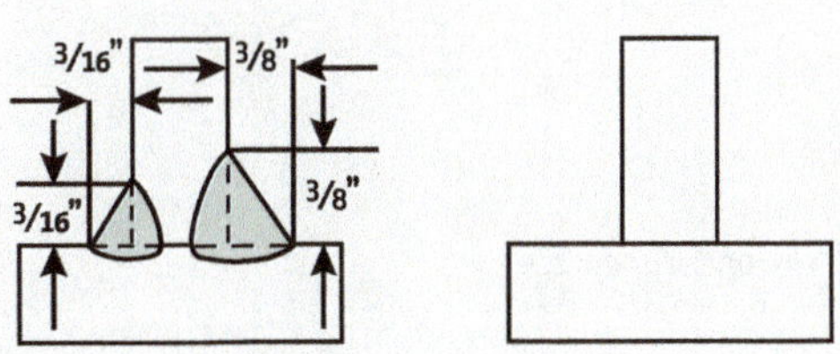

6. 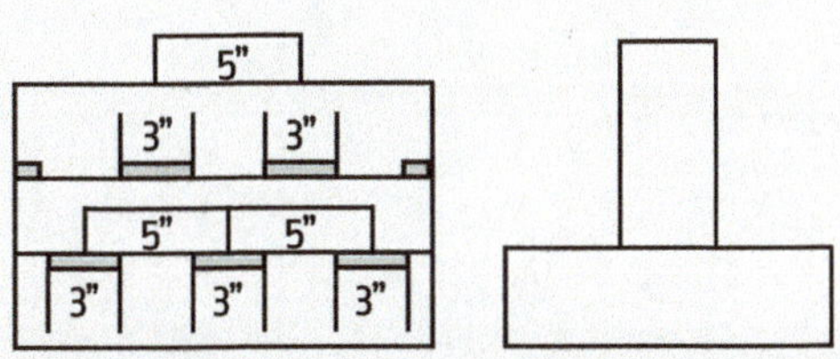

7. 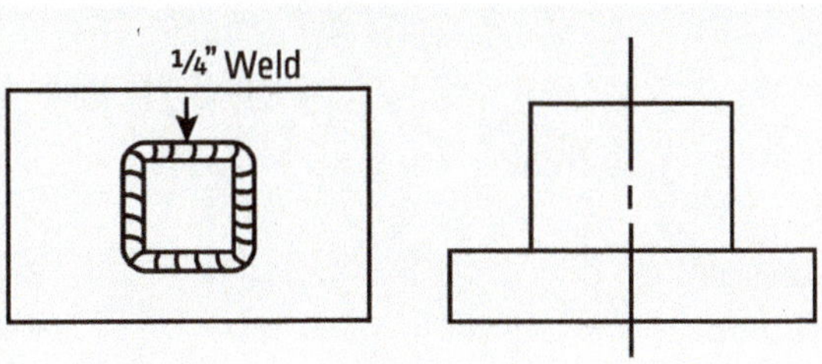

8. 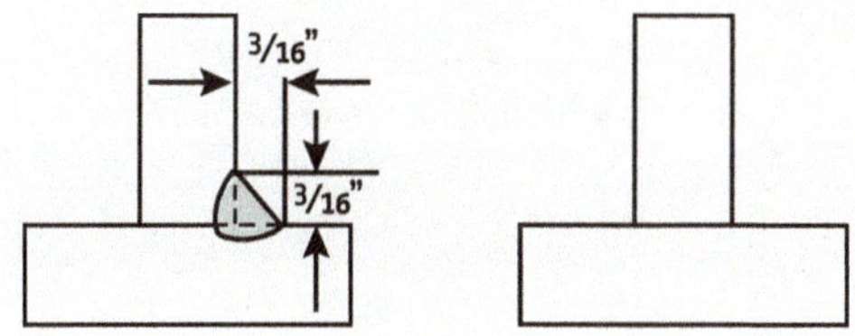

9. 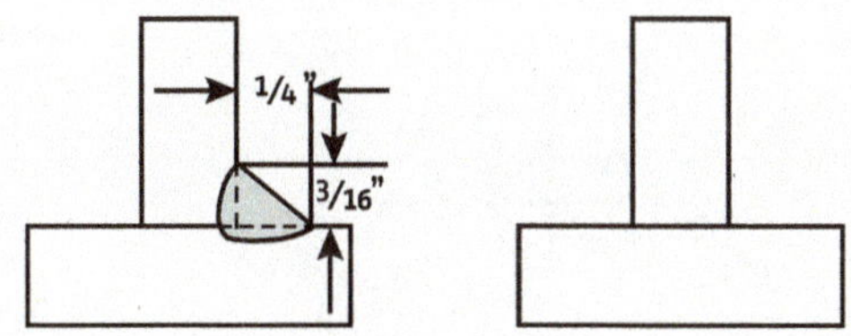

10. 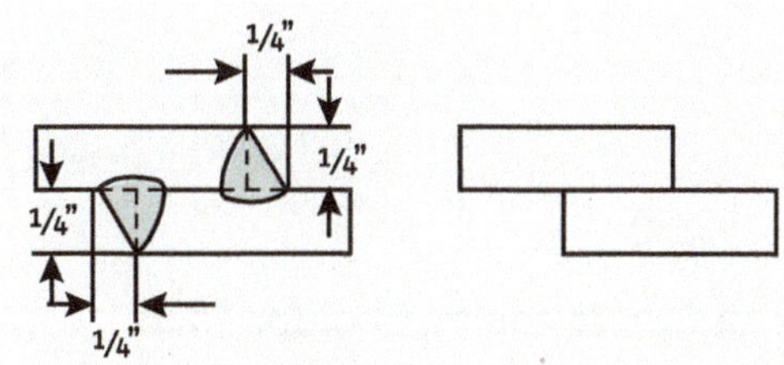

11. 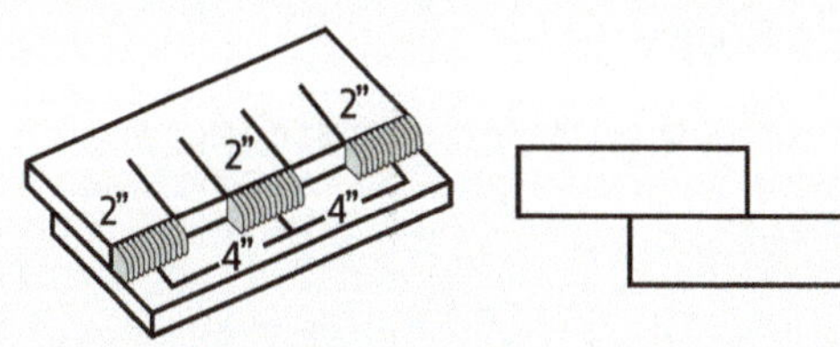

12. 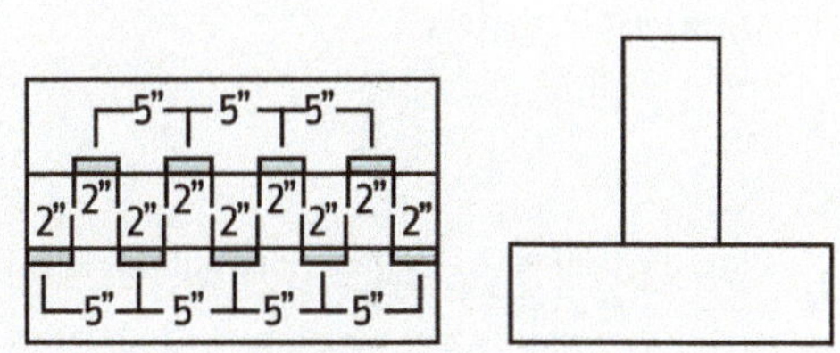

13. 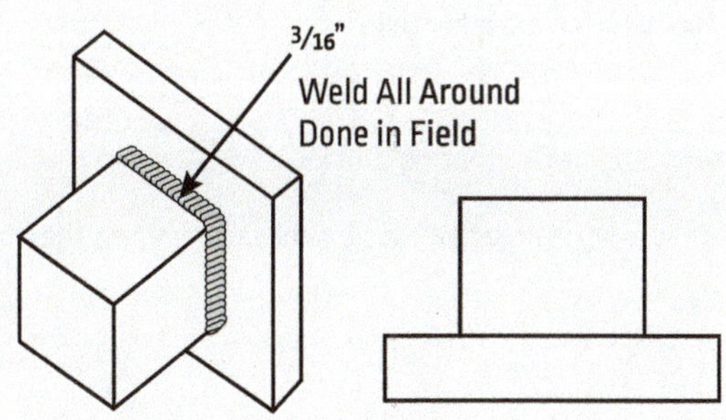

Review Questions

Describe each weld symbol (type, size, side and length).

1. (Example)

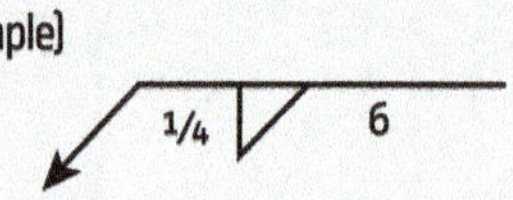

Type: Fillet
Size : 1/4
Side: Arrow Side
Length: 6"

3.

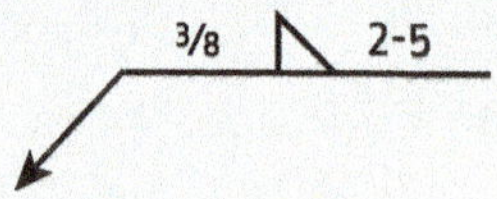

Type:
Size :
Side:
Length:

5.

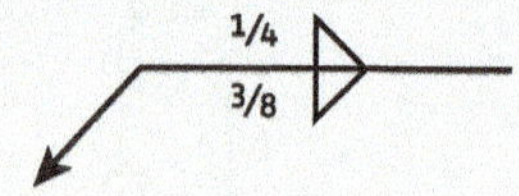

Type:
Size :
Side:
Length:

2.

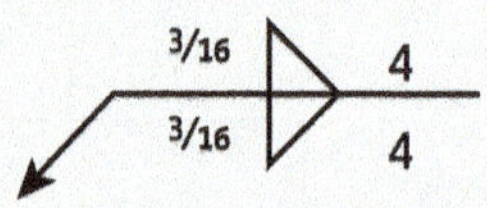

Type:
Size :
Side:
Length:

4.

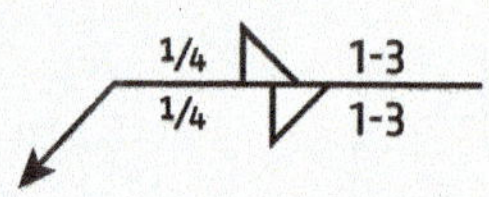

Type:
Size :
Side:
Length:

6.

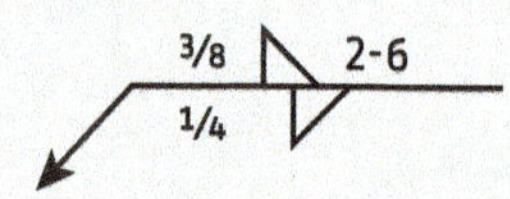

Type:
Size :
Side:
Length:

Sketch and draw the symbol for each:

1.	(Example) A 3/16 in. fillet weld both sides on a lap joint.	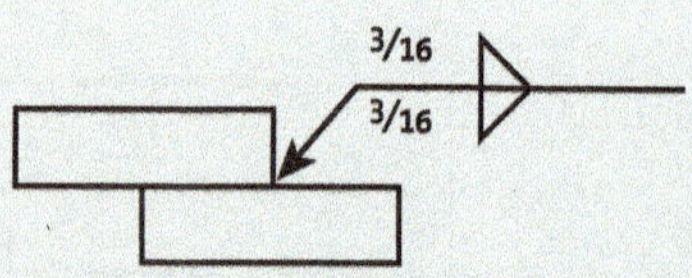
2.	Draw a T-joint with intermittent fillet welds on the arrow side with a weld length of 2 in. on 5 in. centers.	
3.	Draw a T-joint with staggered intermittent fillet welds with a weld length of 1 in. on 3 in. centers on the arrow side.	
4.	Draw a T-joint with fillet welds on both sides and a weld size of 3/16 in. on arrow side and 1/4 in. on the other side.	
5.	Draw a lap joint with a 3/16 in. weld on both sides. The weld is 7 in. long.	
6.	Draw a fillet weld pipe to plate done in the field with a 1/4 in. weld all the way around.	

Notes

Notes

CHAPTER 6: Groove Welds

Chapter Objective

After learning the information in this chapter you will be able to:

- Interpret needed dimensions to prepare groove welds
- Compare groove to other welds
- Understand the components that make up a groove weld
- Identify groove weld depths, bevel angles, groove angles, and root openings
- Identify backing, back welds, backing welds and back gouging
- Identify melt-through, spacers, and consumable inserts
- Understand groove weld dimensions
- Compare the different type of groove welds

Key Terms

Bevel angles
Weld throat
Weld root
Joint
Square groove
Bevel groove
Flare groove
U-groove
J-groove
Intermittent groove
Backing bar
Insert and spacer
Back weld
Backing weld
Back gouging

As we discussed in chapter five, there is a difference between the type of joint and the type of weld.

There are several different types of weld joints that are determined by factors such as the welding process, the thickness and type of materials and the welding position. The welding symbols in Table 6-1 identify all of the different types of groove welds. We can also use a combination of different welding joints to form a groove weld.

Groove welds are formed in a groove between two pieces of material. The edges of the metal that form the groove must be prepared using a joint design such as a square-groove, V-groove, J-groove, bevel-groove, U-groove, flare–V, or flare. Groove welds are generally made on thicker sections of metal.

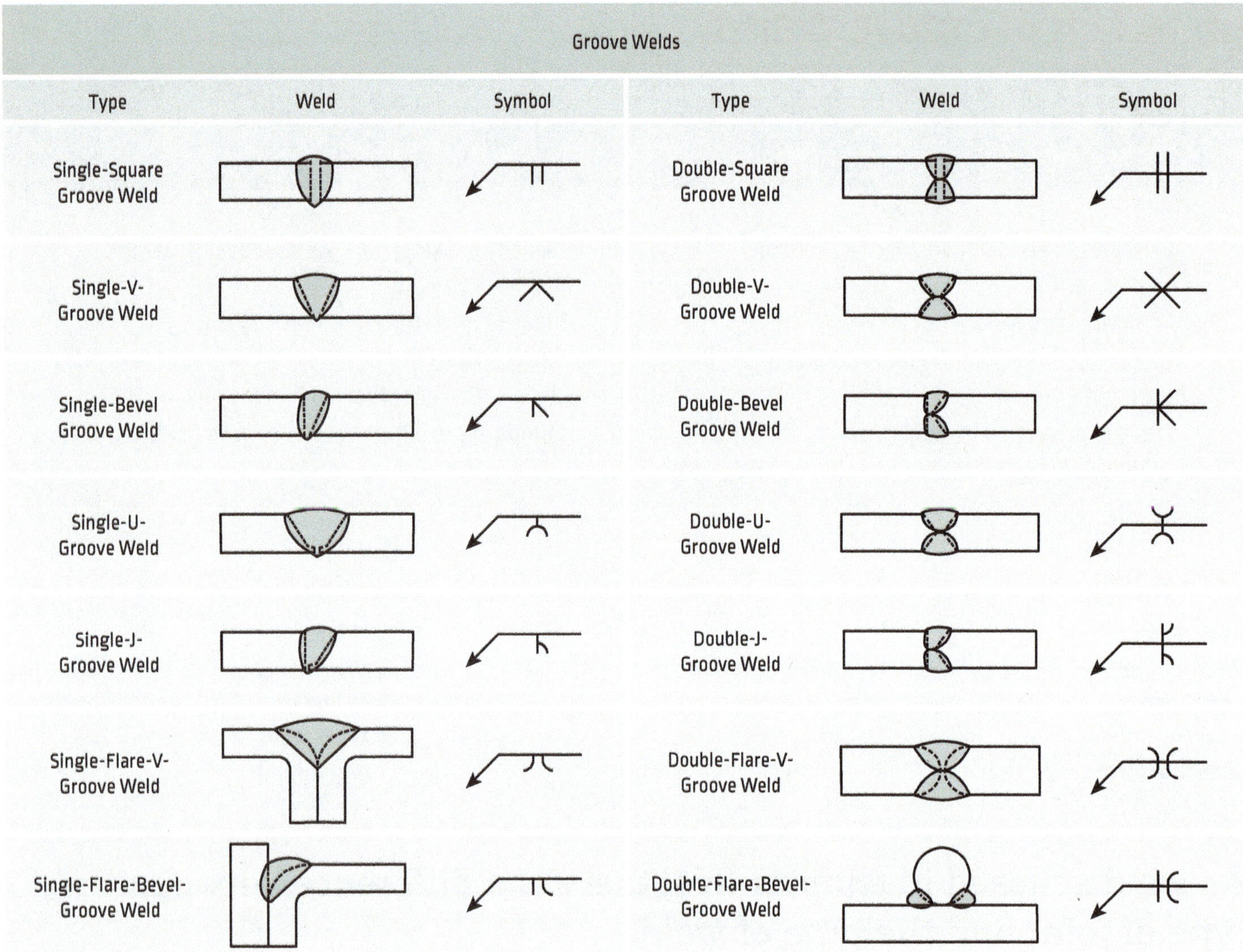

Table 6.1 Groove weld symbols.

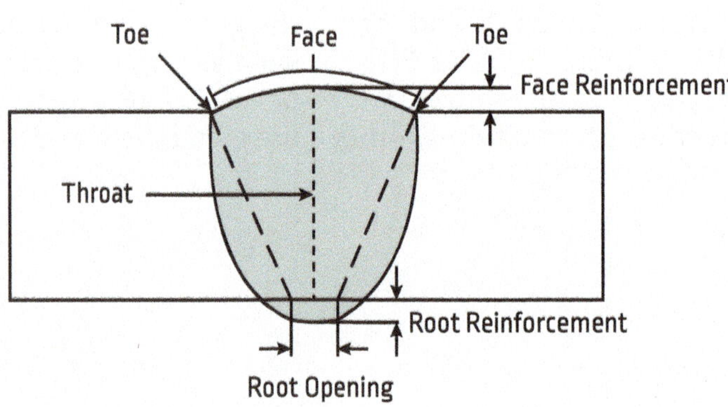

Figure 6.1 Parts of a groove weld.

Square Groove Weld

Square groove welds require the least material preparation of any groove weld. We simply butt the square edges of the metal together. This can be done with a tight gap or a root opening to ensure full penetration. Square grooves are used whenever possible because of their low cost and easy fit-up. The thicker the material, the more difficult it is to obtain full penetration on butt joints.

This is a square groove weld in a butt joint with a 1/16 in. root opening and a 1/2 in. weld size.

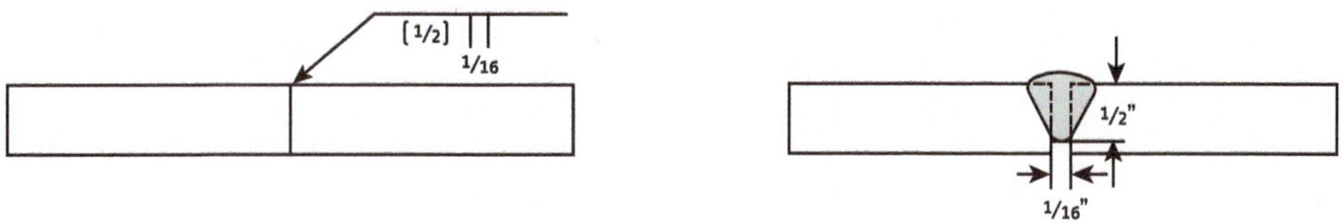

This is a square groove weld in a butt joint with a 1/8 in. root opening, welded from both sides.

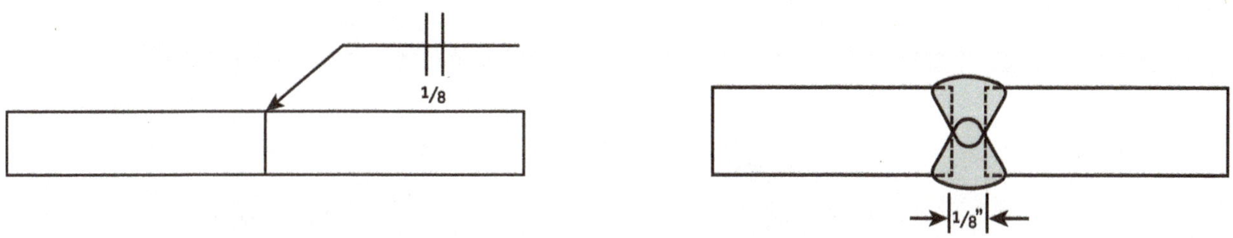

This is a square groove weld in a corner joint with a 3/8 in. weld size.

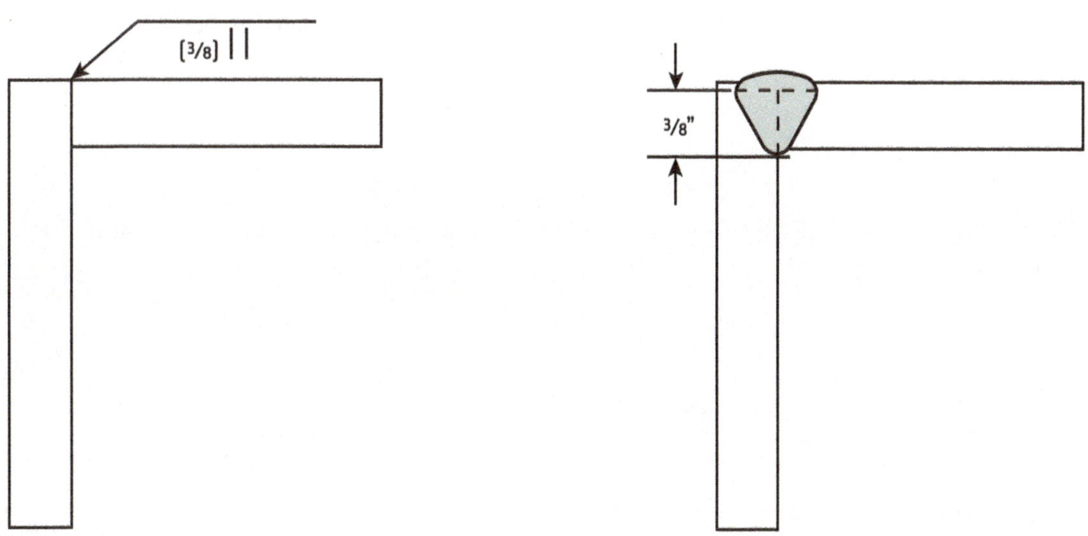

Figure 6.2 Square groove weld.

V-Groove Welds

V-groove welds can be either a single V or a double V depending on the material thickness, process being used or the position of the weld. V-groove welds require that a predetermined angle be cut, ground or machined onto the edge of one or both of the two mating pieces of base material. Single V-grooves or double V-grooves usually require a root opening to ensure full penetration.

This drawing shows a single V-groove weld in a butt joint with a 1/8 in. root opening, 45° included angle with a 3/8 in. depth of groove and full penetration on 1/2 in. material.

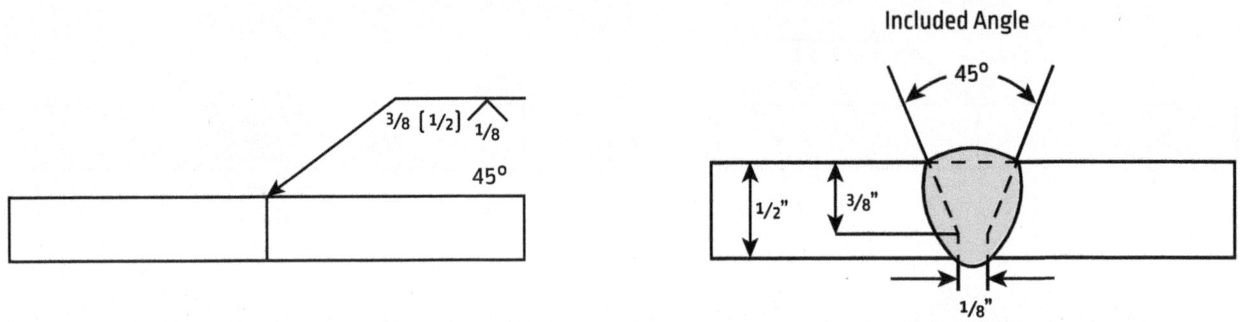

This drawing shows a single V-groove weld in a butt joint on the other side with a 45° included angle and a 1/16 in. root opening.

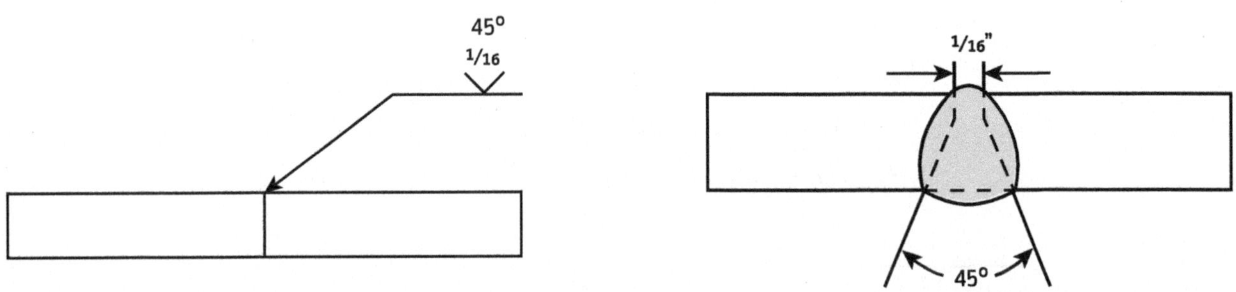

This drawing shows a double V-groove weld. These types of welds are commonly used on thicker materials.

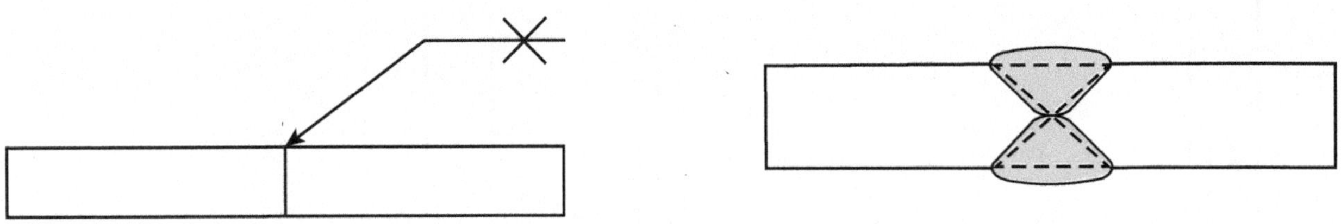

Figure 6.3 Single V-groove welds and double V-groove welds.

Bevel-Groove Welds

Bevel-groove welds can be either a single bevel or a double bevel weld. The print will specify the root opening, bevel angle and groove depth. The symbol for a single bevel will indicate which one of the pieces of base material is to have the bevel. This is important due to stress factors and accessibility to weld the joint.

The drawing shows a single bevel-groove weld. The bevel angle is 35°, the root opening is 1/8 in. with a 3/8 in. groove depth. The drawing also indicates a full penetration weld on 3/4 in. material.

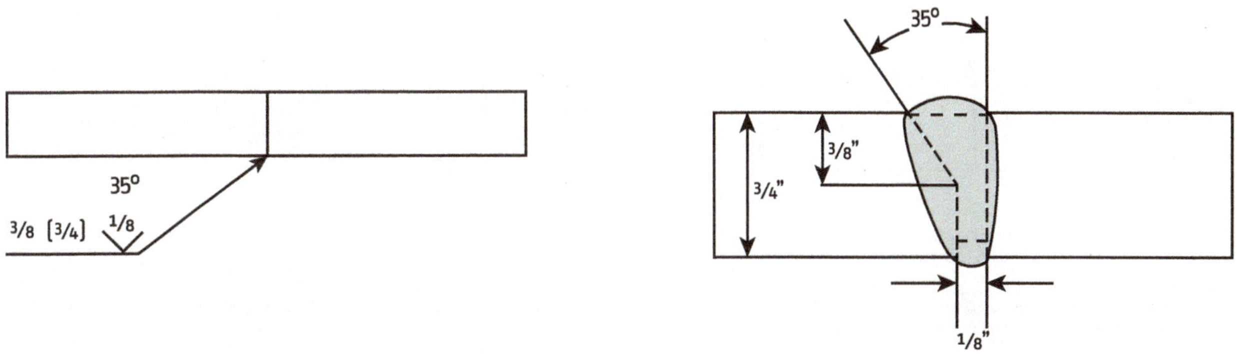

This weld drawing indicates a double bevel-groove weld. The drawing on the right is another example.

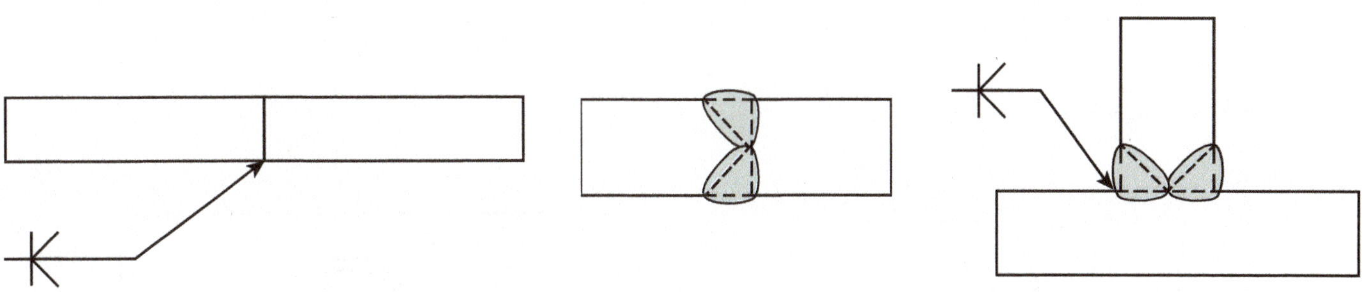

Figure 6.4 Bevel groove welds.

U-Groove and J-Groove Welds

Bevel-groove welds are usually used on materials over 1.0 in. thick. While you could try to weld 1.0 in. thick material from one side using a wide V-groove weld, the number of weld passes and the amount of time required to perform such a weld would be tremendous; distortion would also be likely because of the large heat input. It's both easier and more efficient to weld thicker materials using groove welds such as a U- or J-groove. The U-groove weld is like a modified V-groove weld, and the J-groove weld is like a modified bevel-groove weld. Both of these weld types will allow you to form good welds on thick materials with less weld metal, less distortion and certainly less cost.

U-Grooves

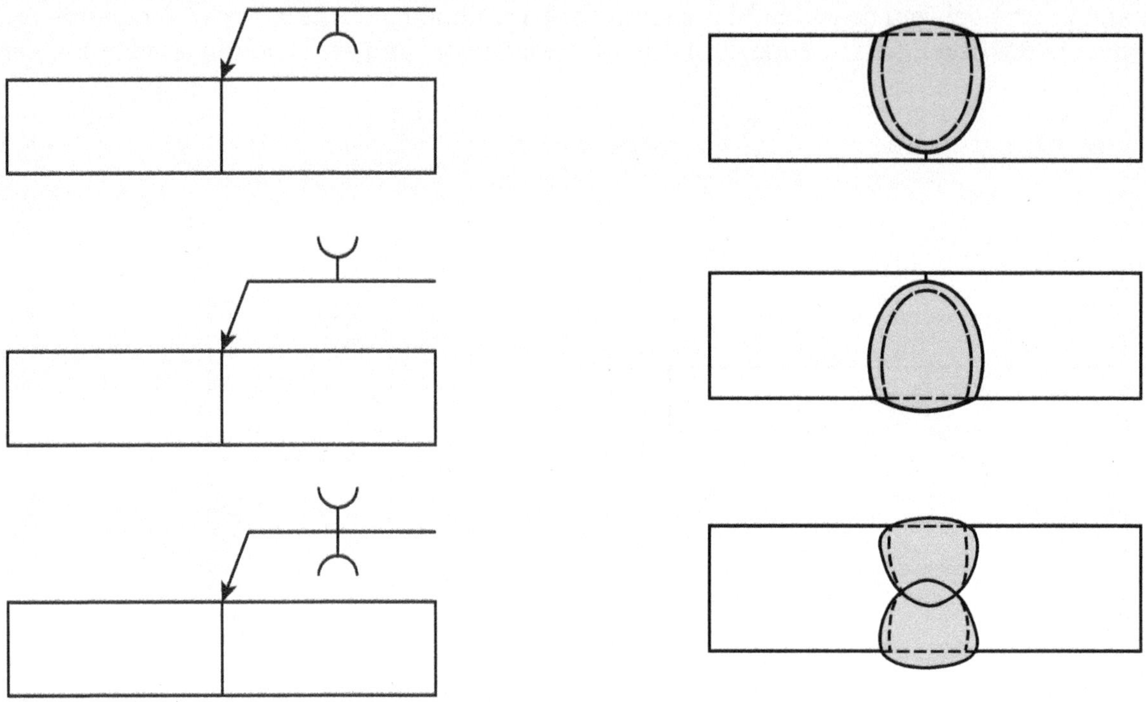

J-Grooves

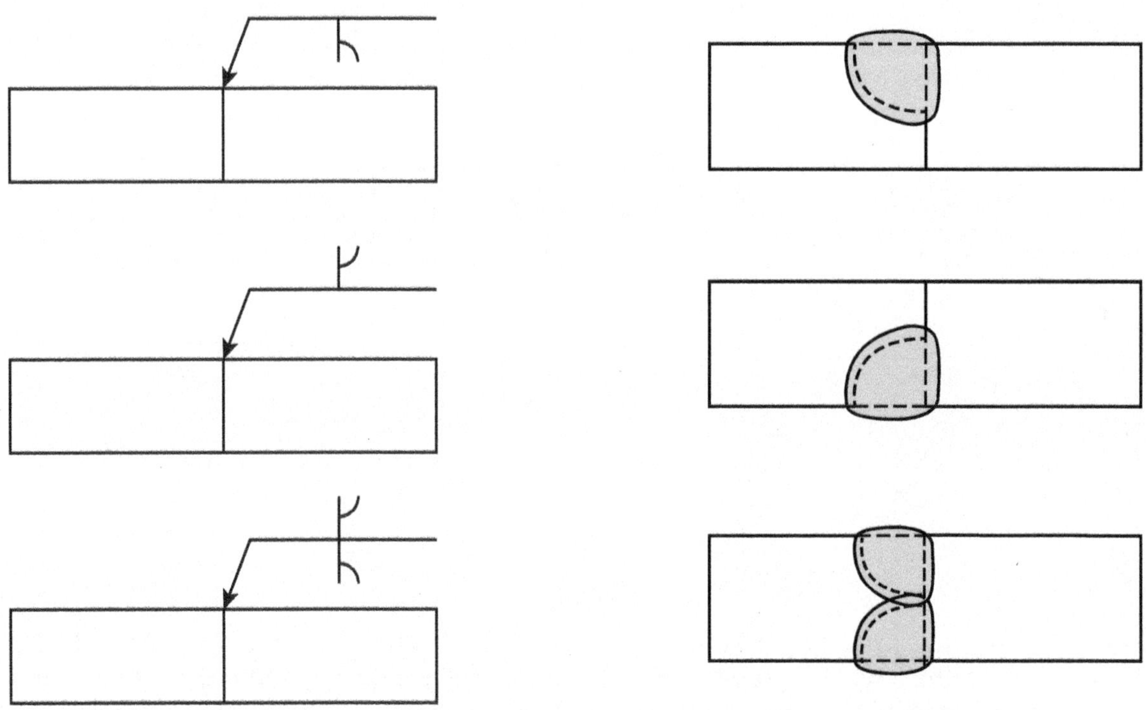

Figure 6.5 J- and U-groove symbols and sketch of that weld type.

Flare V-Groove and Flare Bevel-Groove Welds

Flare V-groove and flare bevel-groove welds are used to weld round or square tubing with rounded corners. An example of this type of joint is welding a round bar that is lying on a flat plate. These weld joints are different from most because you cannot control the configuration of the joint; instead, the size of the weld and the depth of the groove are dependent upon the shape of the materials being welded to each other.

This drawing shows a double flare V-groove weld on both sides.

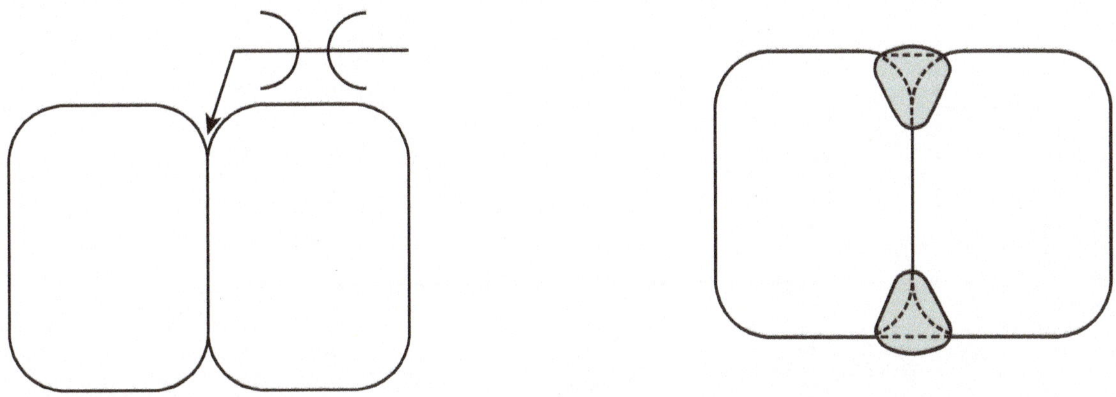

This is the same joint but only welded from one side.

This is a single flare bevel. The drawing shows the rod radius and the groove depth.

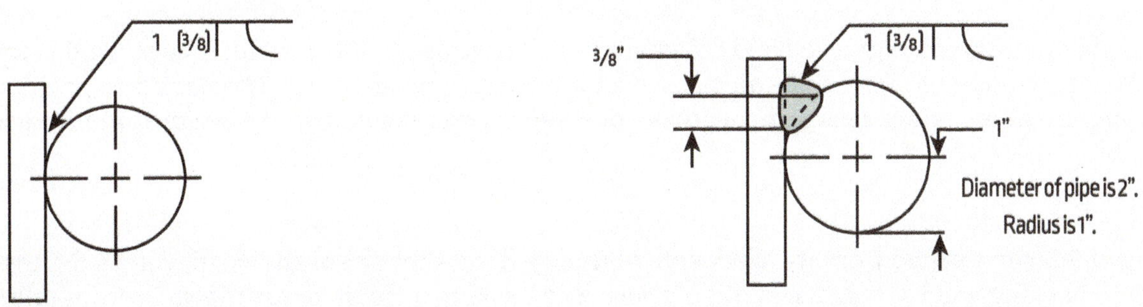

Figure 6.6

In most cases, you cannot obtain full penetration using a flare-type weld joint. It's acceptable to use a root opening if deeper penetration is required. The depth of the groove weld is typically equal to the radius of the bar being welded, as shown in Figure 6.6 (bottom).

Intermittent Groove Welds

Intermittent groove welds are used similarly to intermittent fillet welds, just in a different type of joint. Intermittent groove welds are welds that are not solid all the way across the joint. An intermittent groove weld is indicated with a different symbol than a fillet weld, but the length and pitch measurements are the same. The length of the weld and the pitch (center-to-center) measurements are specified on the left side of the weld symbol (Fig. 6.7).

Groove Welds

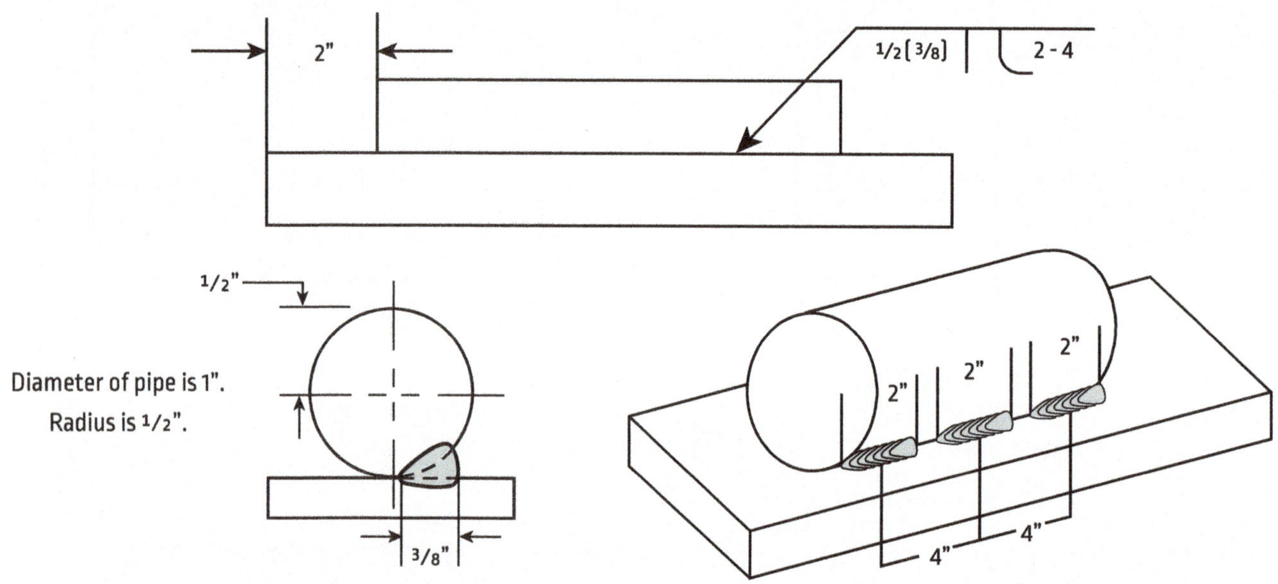

Figure 6.7 Weld symbol and illustrations for a intermittent groove weld.

Intermittent groove welds are commonly used when welding a tube laid flat onto a flat base material or when welding a square tube flat onto another flat piece of base material. We might also see these types of welds used to weld two pieces of round stock butting up to each other in a parallel position.

Back Welds

Back welds are formed on groove welds to ensure full penetration of the weld. A back weld is formed on the back side of a completed weld after all other welding operations are done. The back side of the weld is ground out or air carbon arc gouged, creating a groove deep enough to reach into the weld from its back side.

Backing Welds

Backing welds are formed from the back side of the weld, on the root of the weld, before the rest of the weld is formed. A backing weld serves the same purpose as a backing strip: it keeps the weld from falling through the bottom of the weld joint. The difference between a backing weld and a back weld is when the weld is made. Back welds are done after the weld is formed, and backing welds are done before the weld is formed.

The drawing shows the symbol for a "back" weld. The open half circle indicates a back weld. The back weld is always done after the main weld is completed.

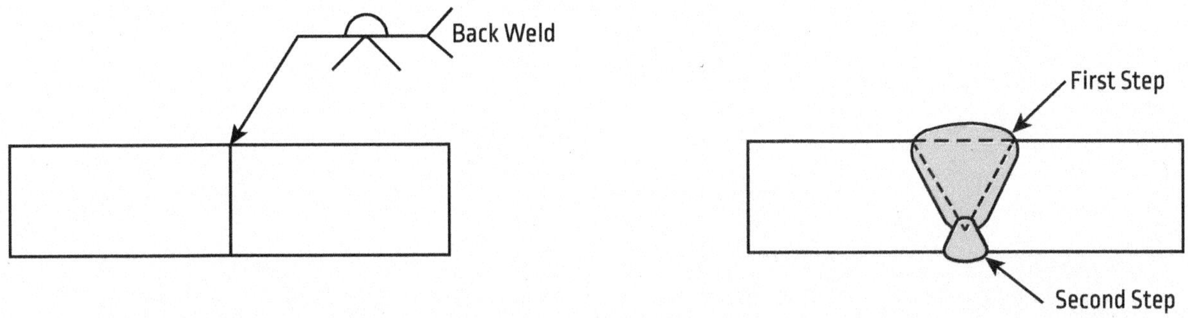

Figure 6.8

The symbol for "back" welds and "backing" welds is simply the order of operations. Back welds are performed after the main weld. Backing welds are performed before the main weld.

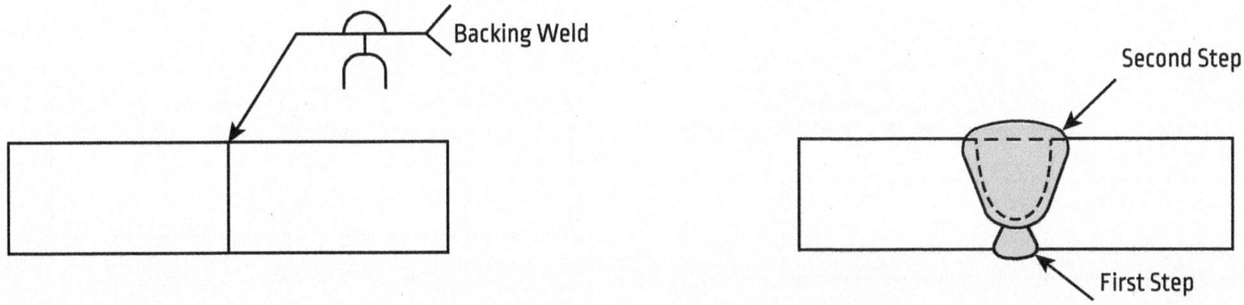

Figure 6.9

Back gouging can be done in several ways. A grinder is a common way of back gouging a weld for preparation of a back welds.

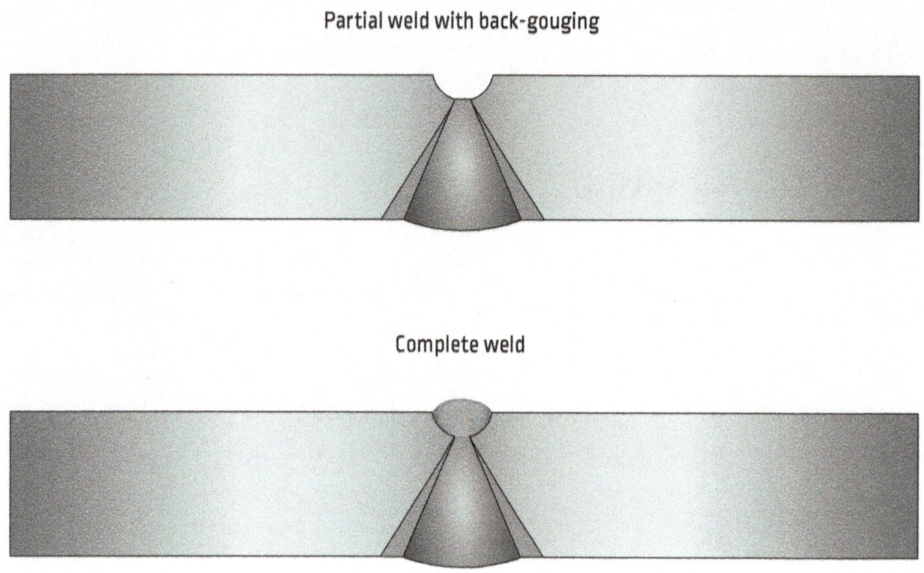

Figure 6.10

Back Gouging

In back gouging, a grinder, carbon arc gouger or plasma gouger is used to remove material from the back of a partially completed weld to ensure fusion of the weld metal placed into the front of the weld joint. Back gouging is usually performed on thicker materials. It can save a great deal of time and labor preparing the base materials. When back gouging is to be performed, it is indicated in the tail section of the welding symbol.

Print	Weld Example

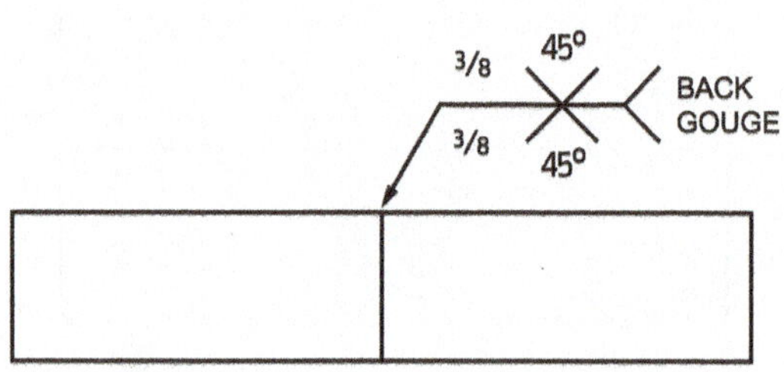

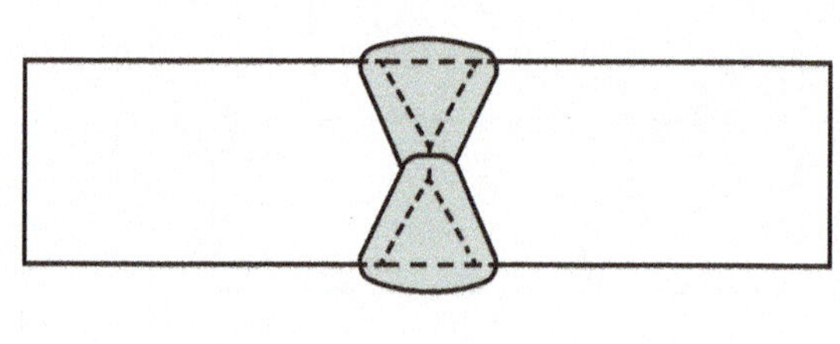

Figure 6.11 Symbol and illustration of weld that is back gouged.

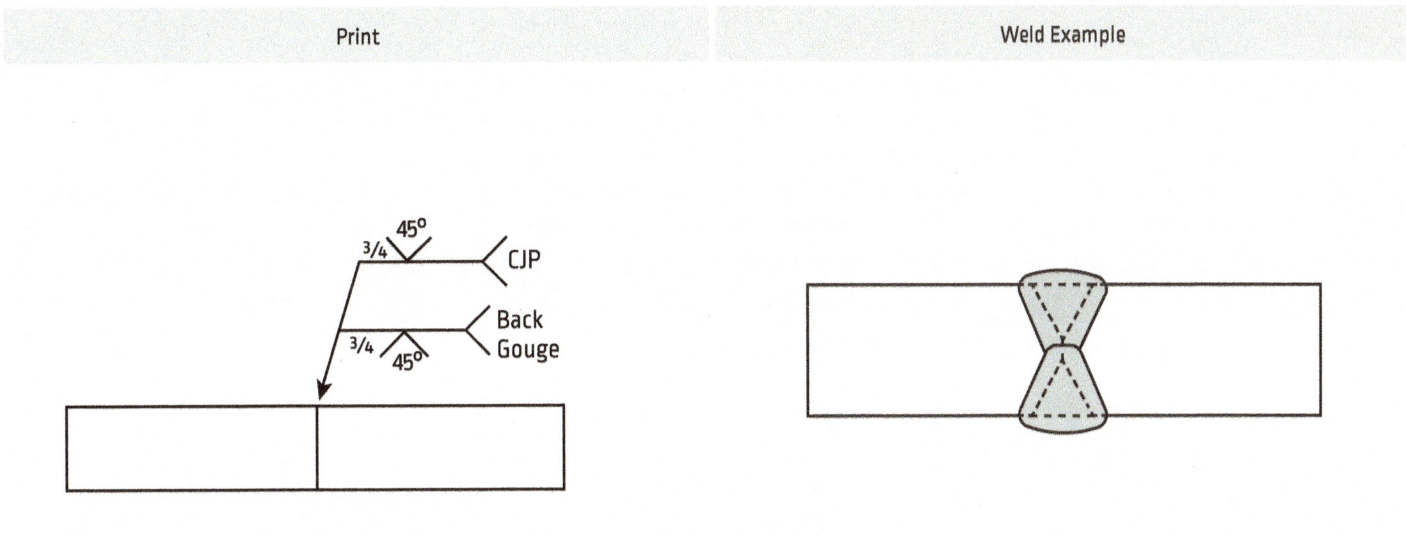

Figure 6.12 Symbol and illustration of weld joint is back gouged.

Backing Bars

A backing bar is used on the root side of the weld to keep the weld from falling through the joint when using a high temperature welding process such as spray transfer GMAW or FCAW. Backing bars are usually removed using an oxyfuel cutting torch with a special scarfing tip, a plasma cutter with a special scarfing tip, or a grinder. They are sometimes removed by machining. The width and thickness of the backing bar will be specified on the print, and the print will also indicate whether it should be removed or stay in place. A backing bar symbol appears on a drawing as a rectangular box on the root side of the weld. A letter R inside the box indicates that the backing bar is to be removed.

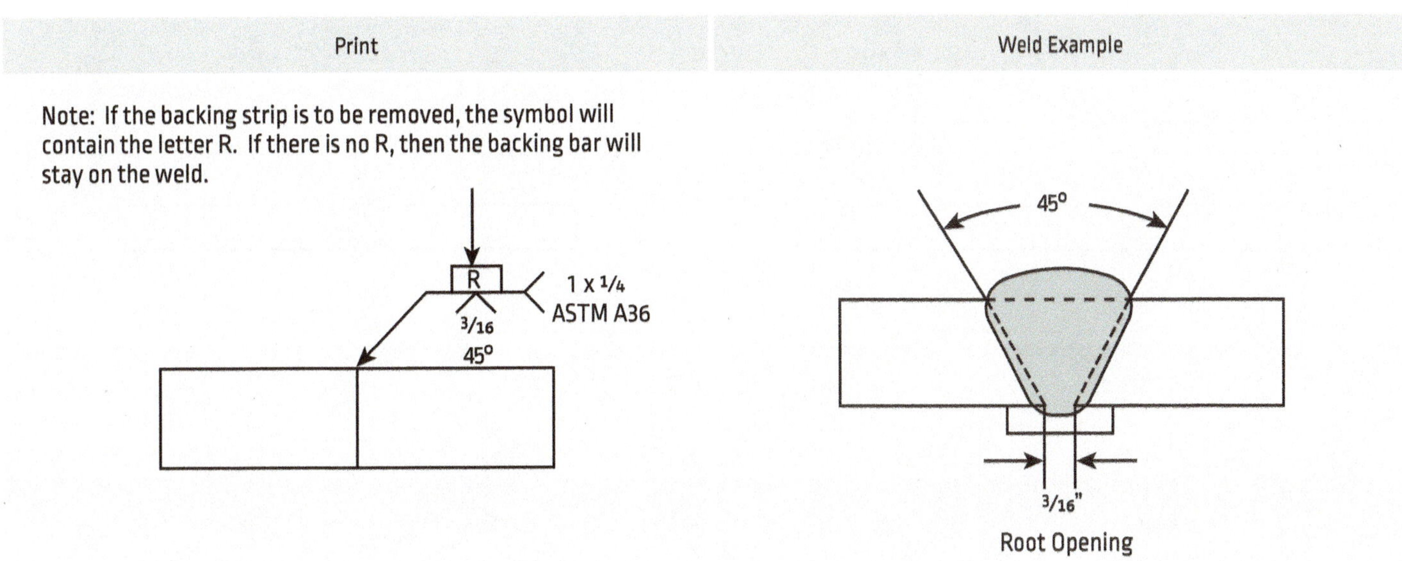

Figure 6.13 Symbol and joint for backing bar.

Spacers

Spacers are used on thicker materials to obtain an even, consistent root opening during fitup. Materials over 1.0 in. thick are generally welded with some type of double groove weld. When the joint is welded from the front, the spacer is consumed as part of the weld. After the first weld is completed, the spacer is ground or gouged out from the back side. After the spacer is removed, the joint is welded from the other side to ensure complete fusion. The symbol for a spacer is a rectangular box, similar to the backing bar symbol, but the box is placed in the middle of the reference line instead of on one side of the reference line (Fig. 6.14).

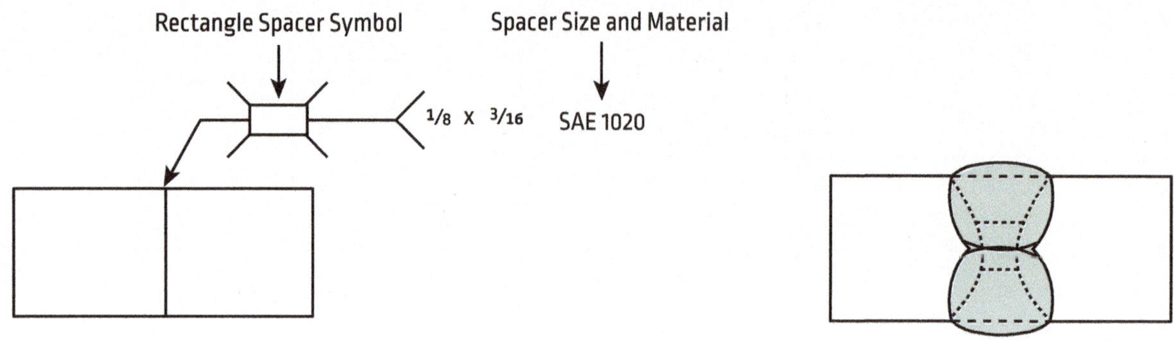

Figure 6.14 Symbol and illustration of spacer.

Consumable Inserts

A consumable insert is used to improve fitup. The difference between a spacer and a consumable insert is that the consumable insert is fully consumed as part of the weld. This means the insert must be made from material compatible with the base material being used. The AWS has specific classifications for consumable insert types, which are shown in the tail section of the welding symbol (Fig. 6.15). The consumable insert symbol is a square on the root side of the reference line.

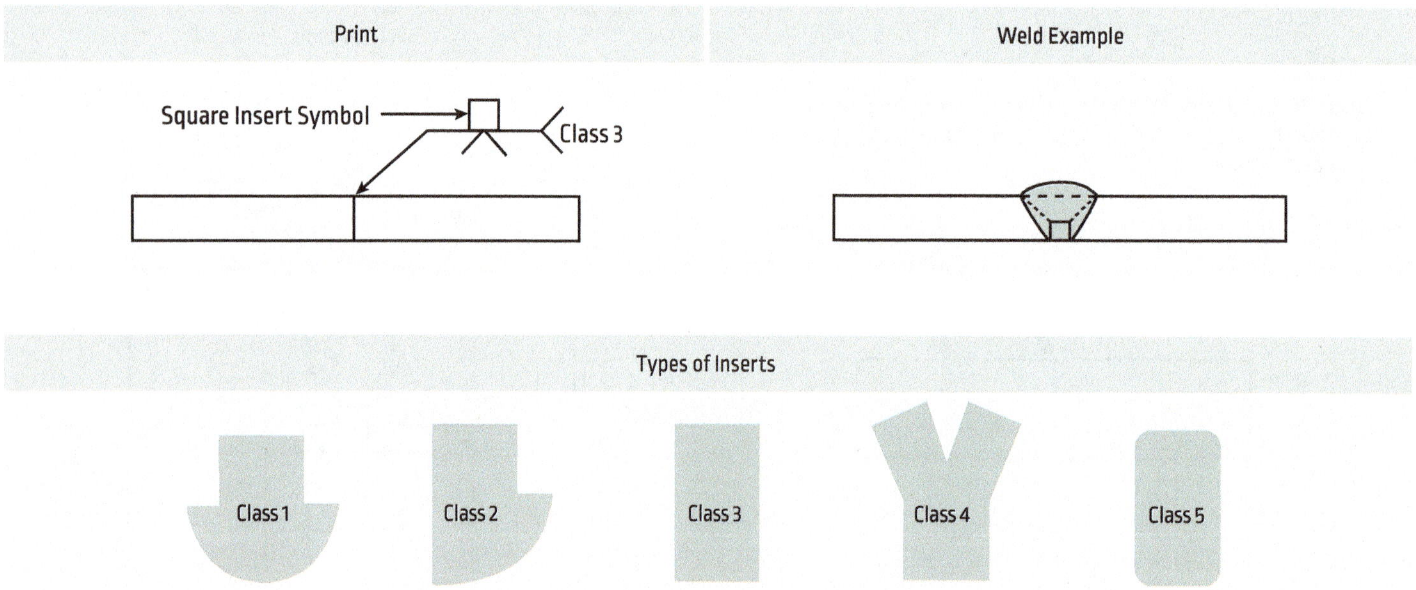

Figure 6.15 Consumable insert symbols and their classifications.

Review Questions

Name and sketch the weld that the symbol represents.

1. Example

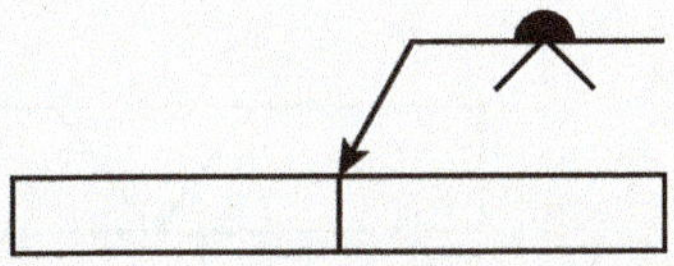

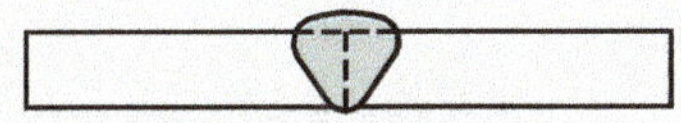

Joint Type: V-groove with melt-through

2.

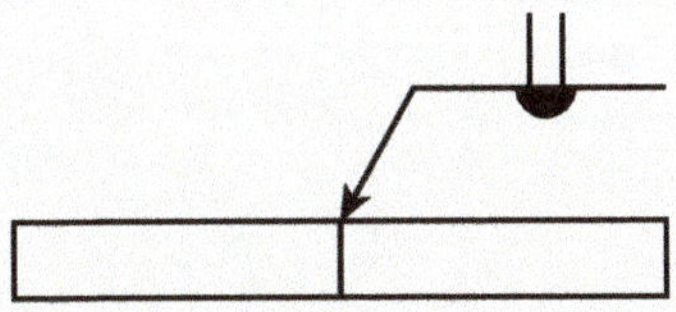

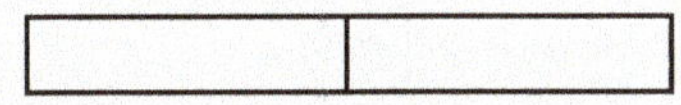

Joint Type:

3.

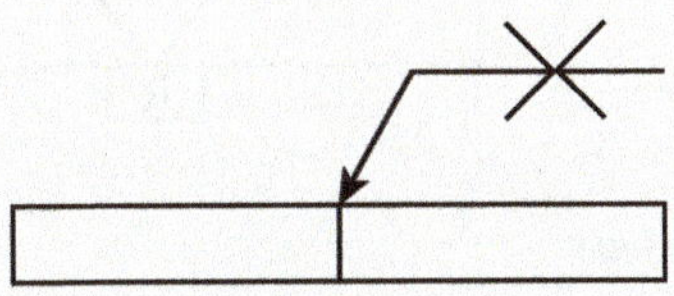

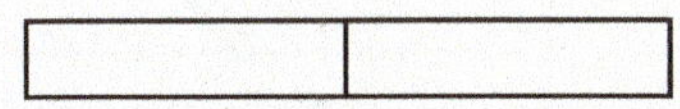

Joint Type:

4.

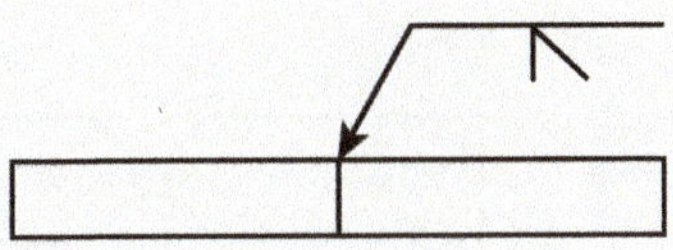

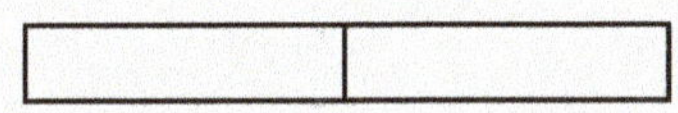

Joint Type:

5.

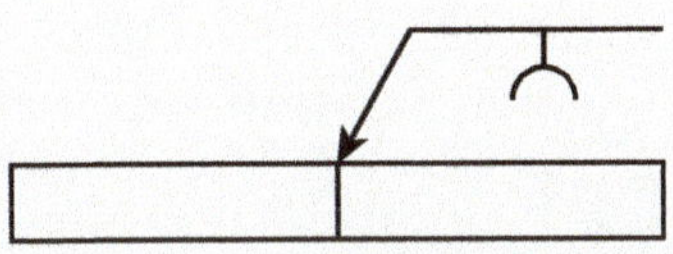

Joint Type:

6.

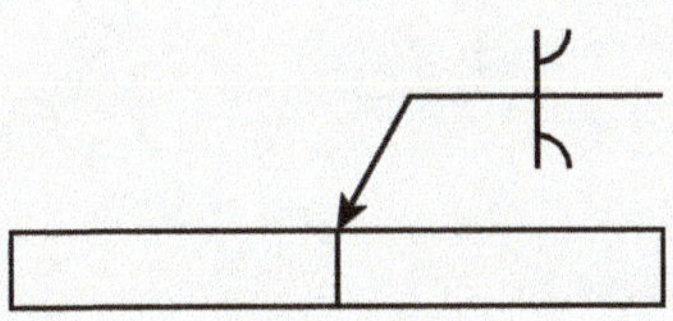

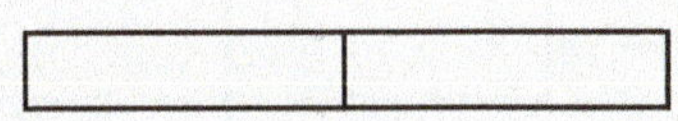

Joint Type:

7.

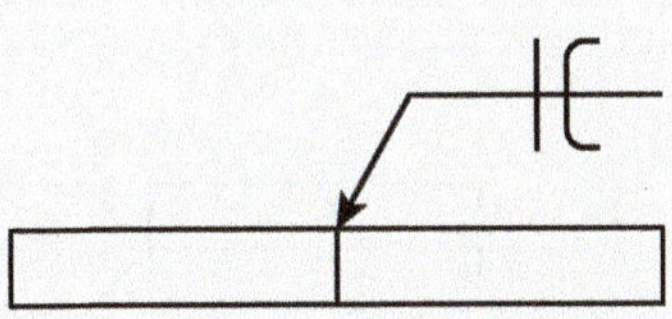

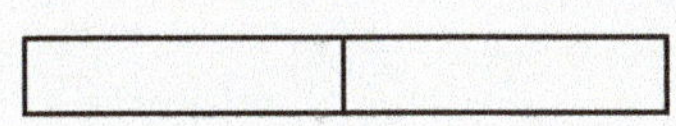

Joint Type:

8.

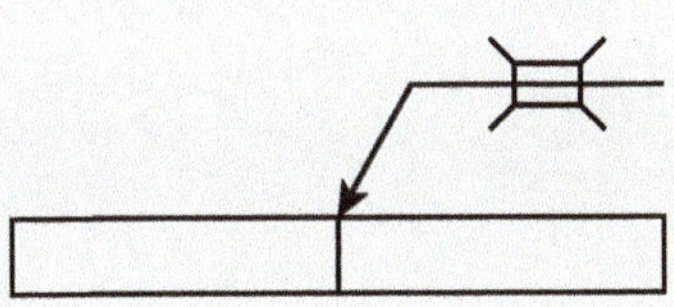

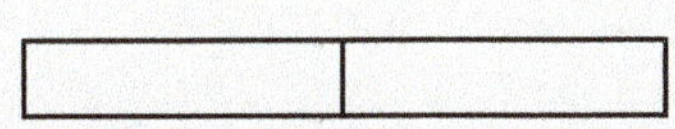

Joint Type:

9.

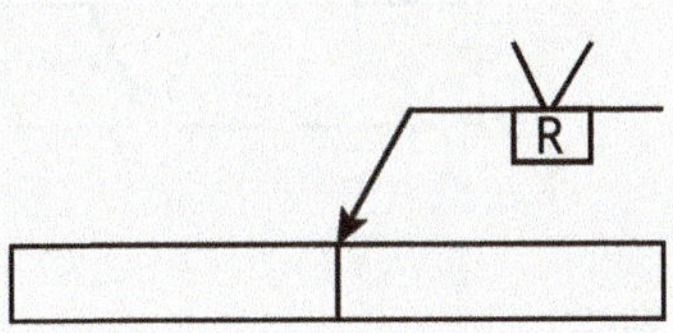

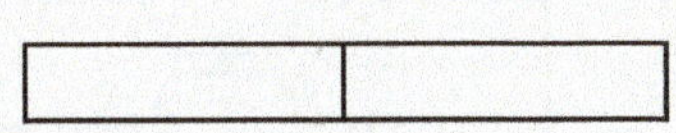

Joint Type:

Review Questions

Name the type of groove weld.

Example

1. 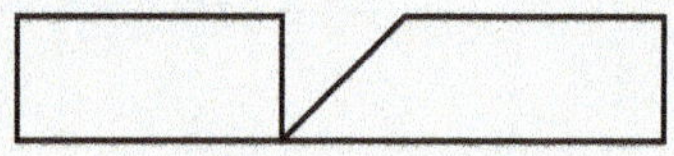

Weld Type: Single bevel weld

2. 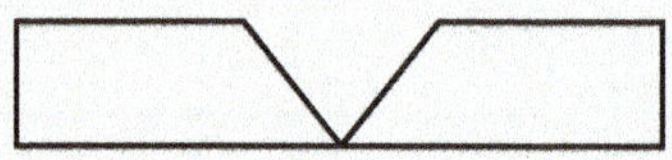

Weld Type: ______

3. 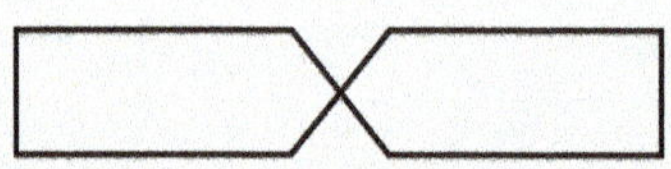

Weld Type: ______

4. 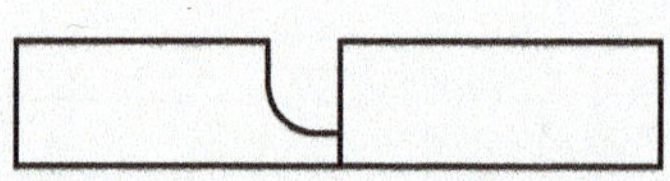

Weld Type: ______

5. 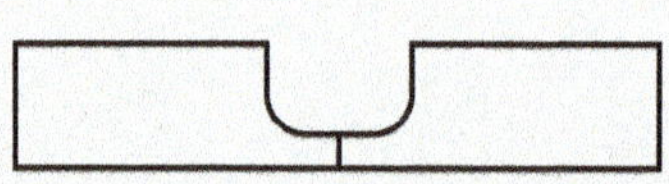

Weld Type: ______

6. 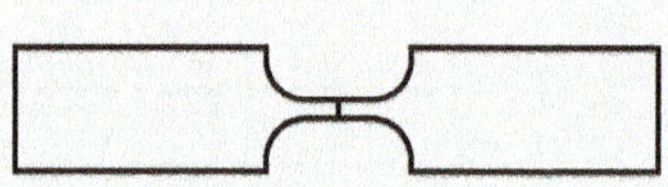

Weld Type: ______

7. 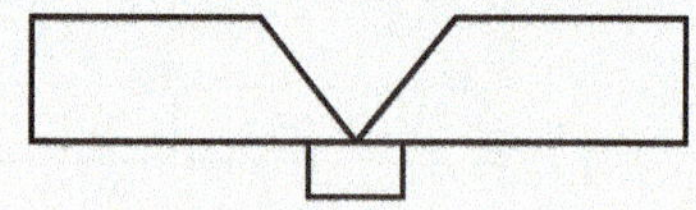

Weld Type: ______

8. 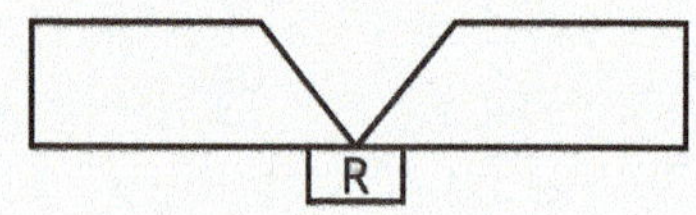

Weld Type: ______

9. 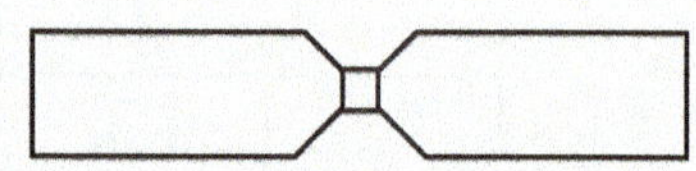

Weld Type: ______

10. 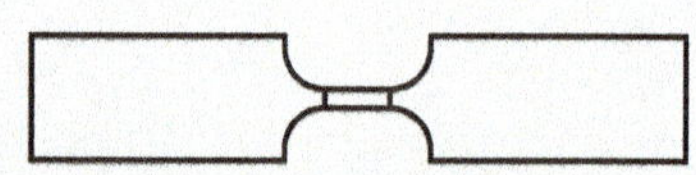

Weld Type: ______

11. 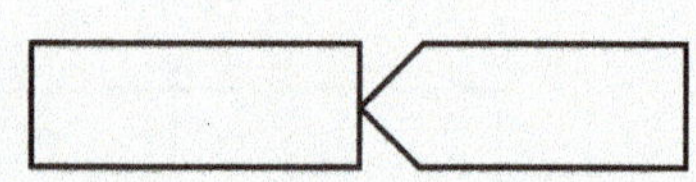

Weld Type: ______

12. 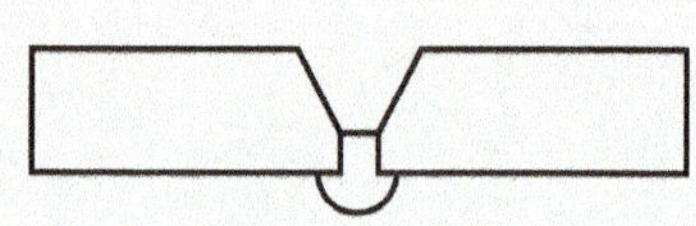

Weld Type: ______

Notes

Notes

CHAPTER 7: Plug, Slot and Edge Welds

Chapter Objective

After learning the information in this chapter you will be able to:

- Understand the difference between plug, slot and edge welds
- Identify strength and size requirements
- Identify weld depth and finishing requirements
- Determine spacing techniques and measurements
- Identify the different symbols for plug, slot and edge welds

Key Terms

Plug welds
Slot welds
Edge welds
Countersinks
Chamfering
Sizing
Strength requirements
Complete joint penetration (CJP)
Partial joint penetration (PJP)

Plug Welds

Plug welds and slot welds are similar, but with one important difference: a plug weld is round and a slot weld is oval-shaped. Plug and slot welds are used when the two pieces of base material overlap. A round or elongated hole is drilled, cut or punched into one piece of base material so that we can form a weld through the hole, fusing the two pieces together. The hole may be completely filled or partially filled as indicated on the welding symbol. Plug weld symbols always provide the angle of countersink, the diameter of the hole and the depth of the weld.

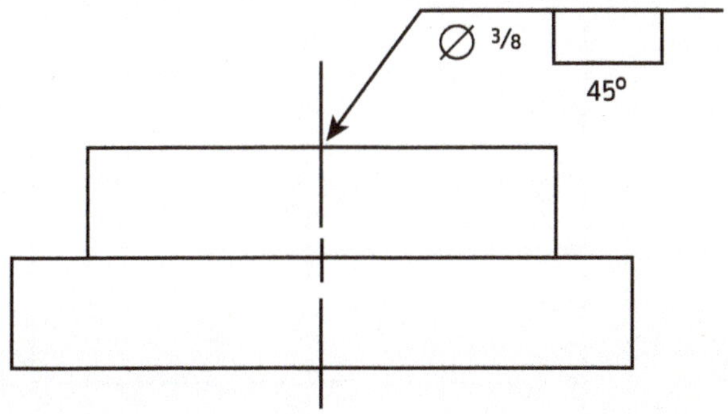

Figure 7.1 Weld symbol for 3/8 in. plug weld with 45° angle of countersink

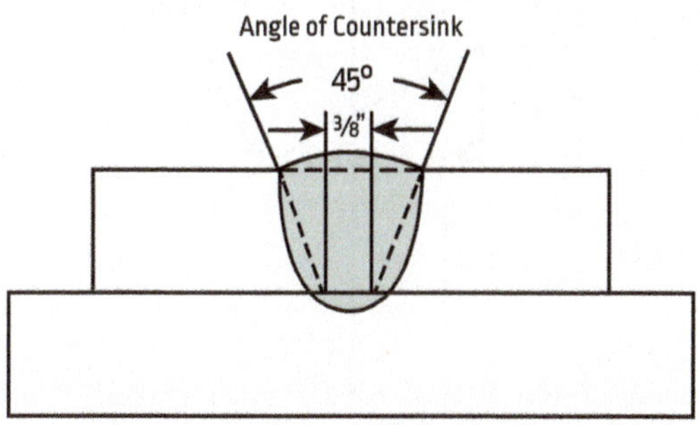

Figure 7.2 Sketch of weld for 3/8 in. plug weld with 45° angle of countersink

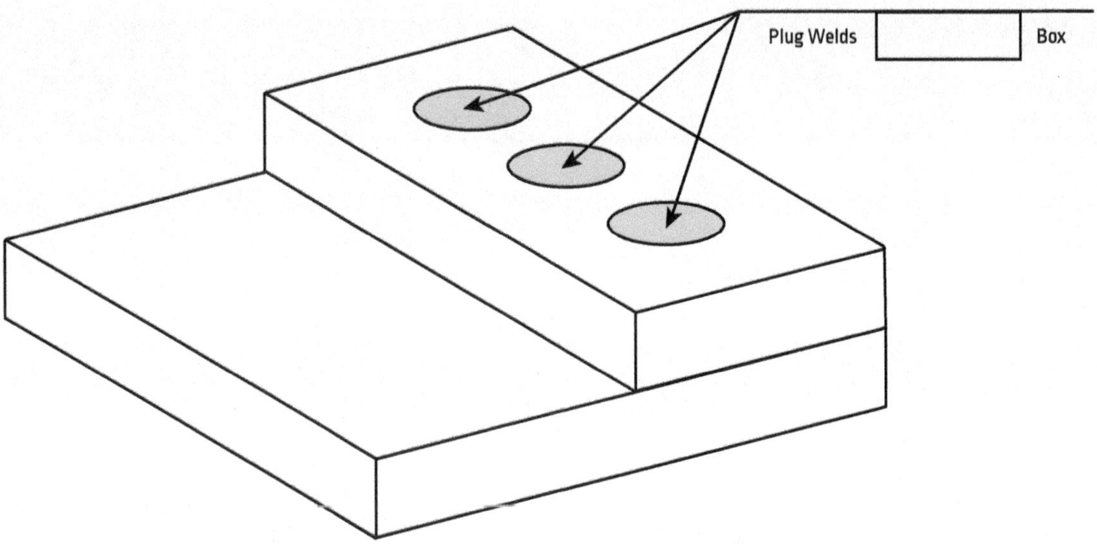

Figure 7.3

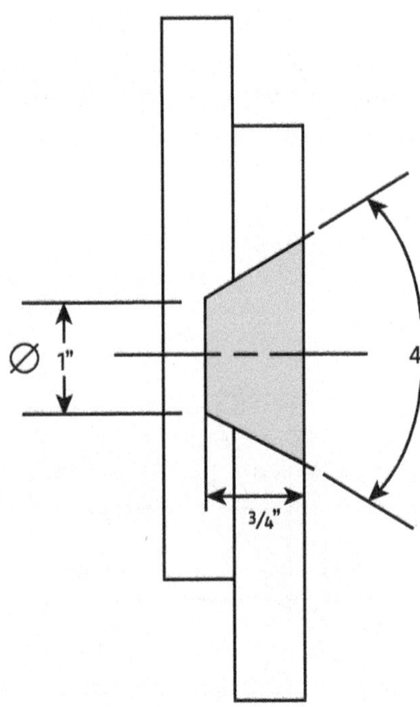

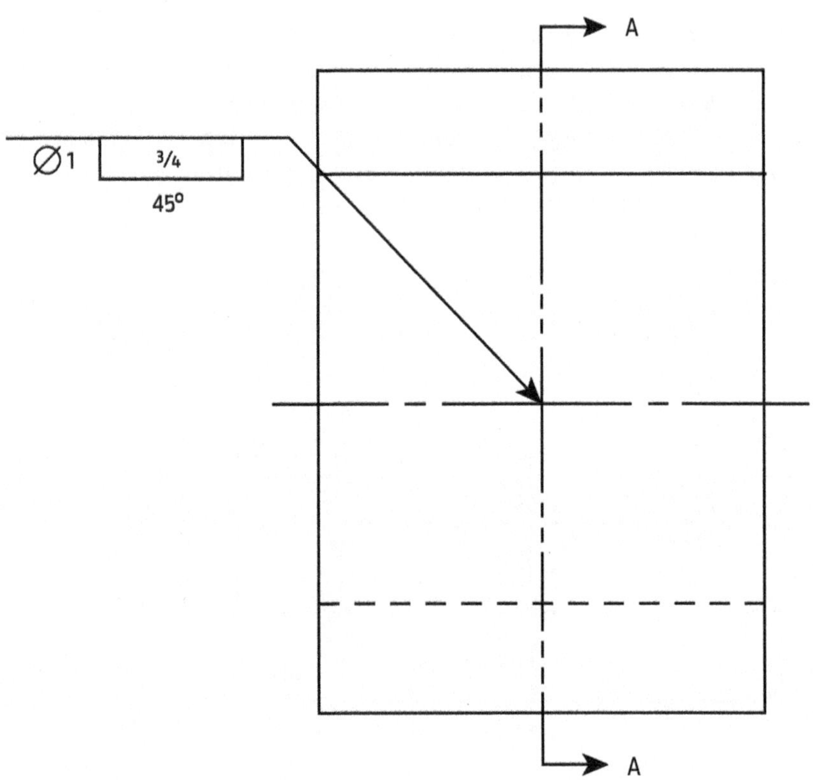

Figure 7.4 Sketch of a 1 in. plug weld with 45° angle of countersink and 3/4 in. depth of filling.

Figure 7.5

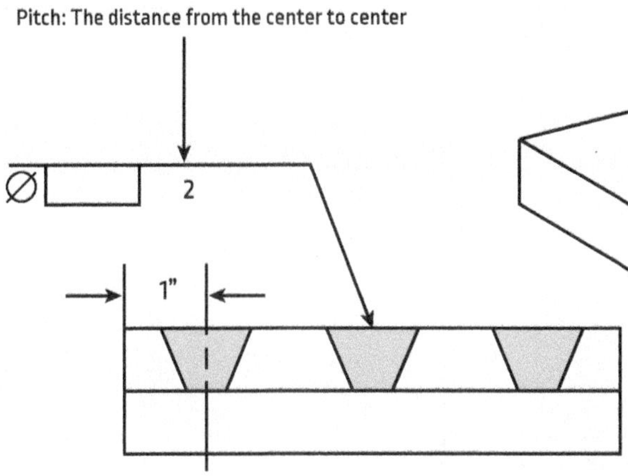

Figure 7.6 A series of plug welds are measured on centers - the same as an intermittent fillet weld.

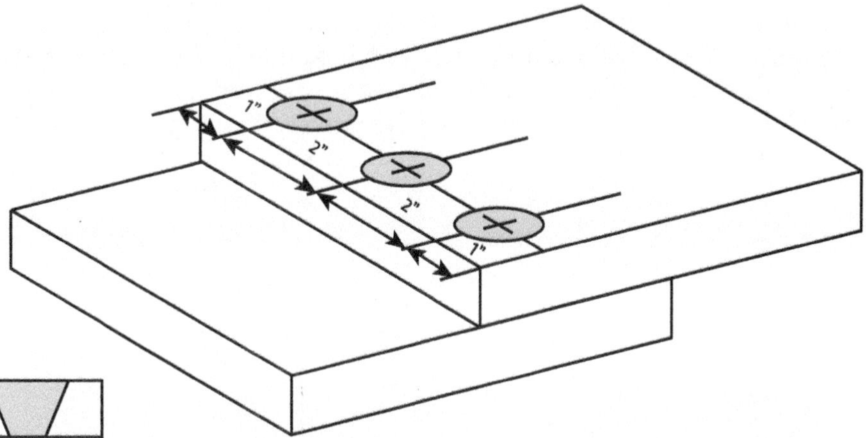

Figure 7.7 Sketch of a series of plug welds with 2 in. pitch

A plug weld is identified as a box on the welding symbol, which can be confused with the symbol for a backing bar. The difference is that a plug weld will have a plug weld symbol to the left of the box, whereas a backing bar will be indicated by a groove weld symbol in addition to the box. Plug welds may also have the edges countersunk. This requires the edges of the holes to be beveled. This will be shown on a drawing with a degree number just outside the box as seen in Figures 7.9 and 7.10.

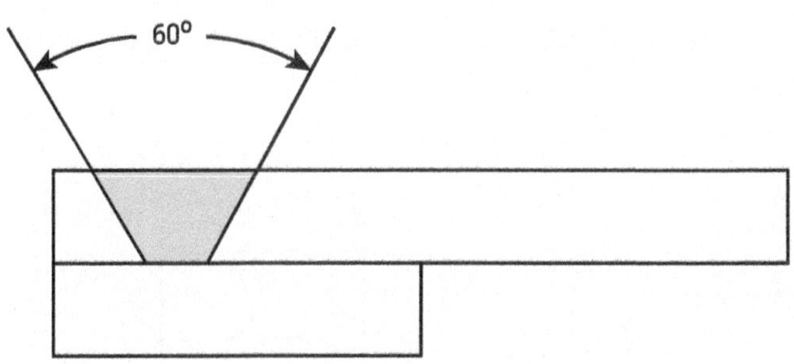

Figure 7.8 Plug weld with 60° angle of countersink

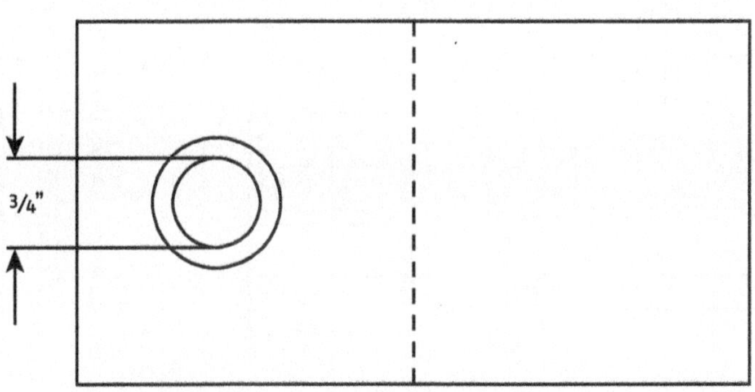

Figure 7.9 Plug weld size of plug weld shown in Figure 7.8

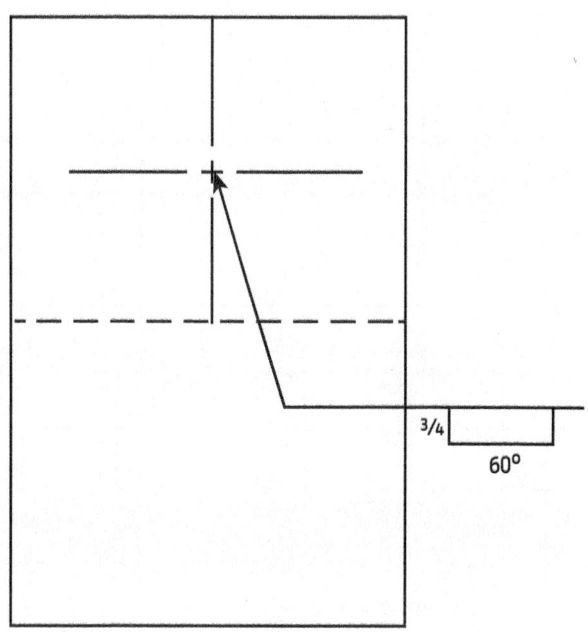

Figure 7.10 Weld symbol for 3/4 in. plug weld with 60° angle of countersink shown in Figures 7.8 and 7.9.

Slot Welds

The slot weld symbol looks similar to the plug weld symbol, but with some important differences. On the welding symbol a slot weld is indicated by a box, like a plug weld. A big difference is that the slot weld has two measurements on the weld symbol. The measurement on the left side of the symbol is the width of the slot, and the number on the right of the symbol is the length of the slot (Fig. 7.12).

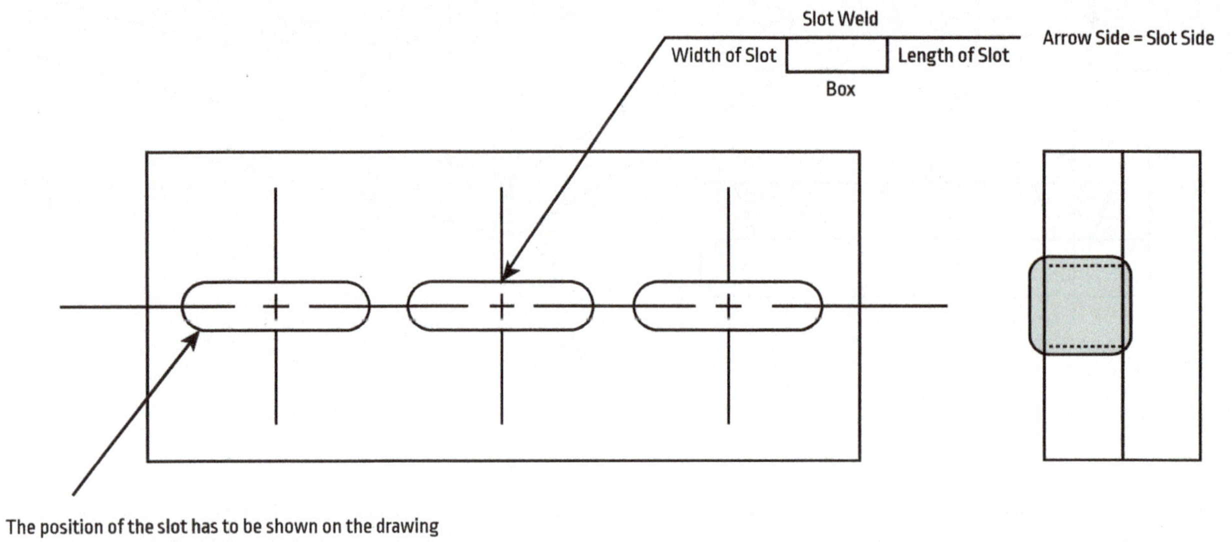

Figure 7.11 Left, plug weld symbol. Right, Sketch of plug weld indicated by symbol.

Multiple slot welds are indicated the same way as plug welds. The pitch is measured from center-to-center and is indicated to the right of the length measurement, which is to the right of the weld symbol (Fig. 7.12). When a slot weld is countersunk, it is indicated the same way as a countersunk plug weld: with a degreed measurement (Fig. 7.13). The thickness of the slot weld, if not filled completely, will be written inside the rectangular weld symbol.

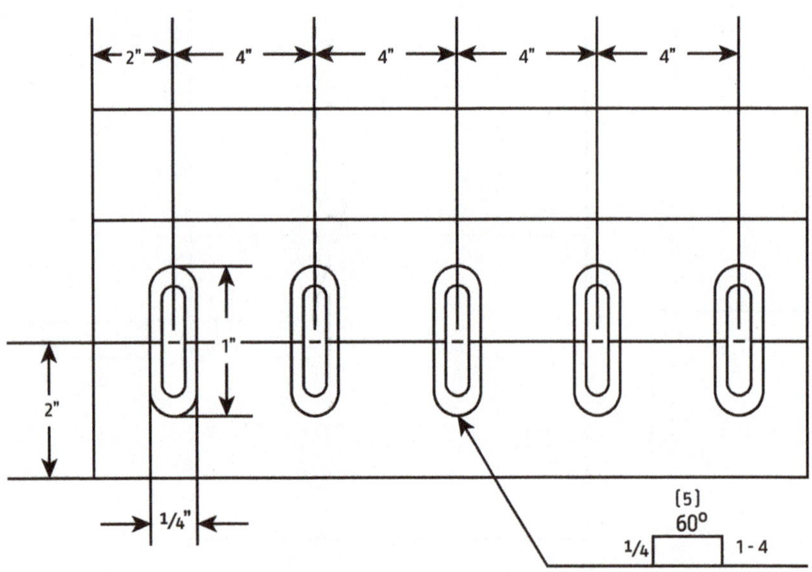

Figure 7.12 Five slot welds - 1/4 in. x 1 in. on 4 in. centers with 60° angle of countersink.

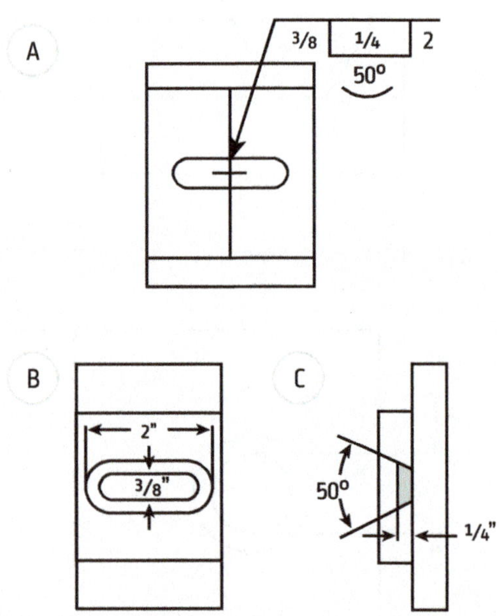

Figure 7.13 The drawing indicates the slot is not to be filled completely. The 1/4 in. inside the slot symbol indicates the depth of filling is 1/4 in.

Edge Welds

Edge welds are formed on the edges of two or more pieces of material configured to create an edge joint. Edge welds are also used with flanged joints and flanged corner joints. Edge welds must fuse all members together with adequate penetration as determined by the designer.

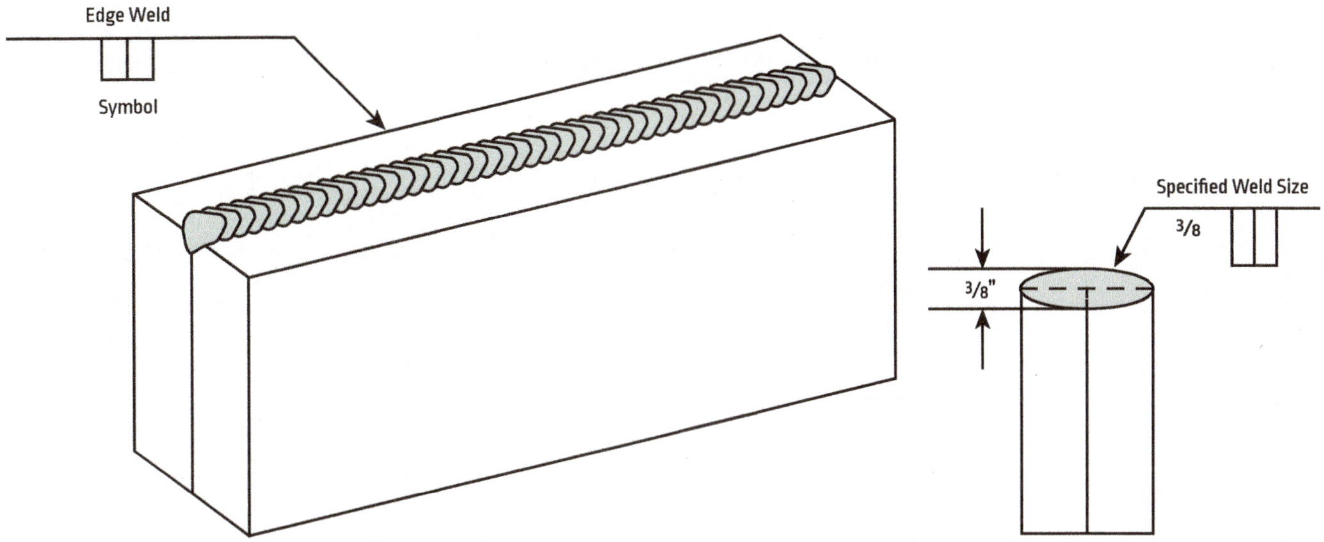

Figure 7.14 The edge weld symbol looks similar to a square groove butt joint symbol except that the ends are closed. If the weld size is not specified, the weld is to fuse the entire thickness of the members.

Several members are often joined with an edge weld. As many as three or four members with mating surfaces might be fused together on the edges of the materials, as shown in Figure 7.15.

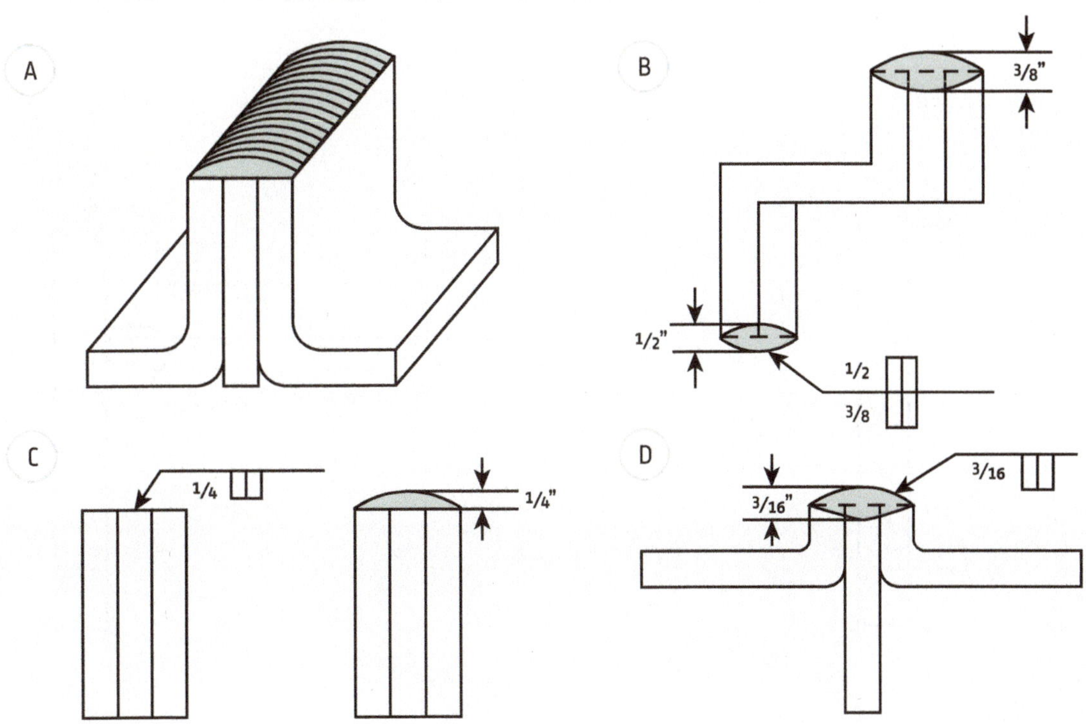

Figure 7.15 Edge welds are sometimes made up of many different members, not just two. A, Example of edge weld with three members. B, 1/4 in. edge weld symbol and weld cross section. C, Material with edge welds of different lengths, cross section and weld symbol. D, 3/16 in. edge weld symbol and cross section.

Full Penetration Edge Welds

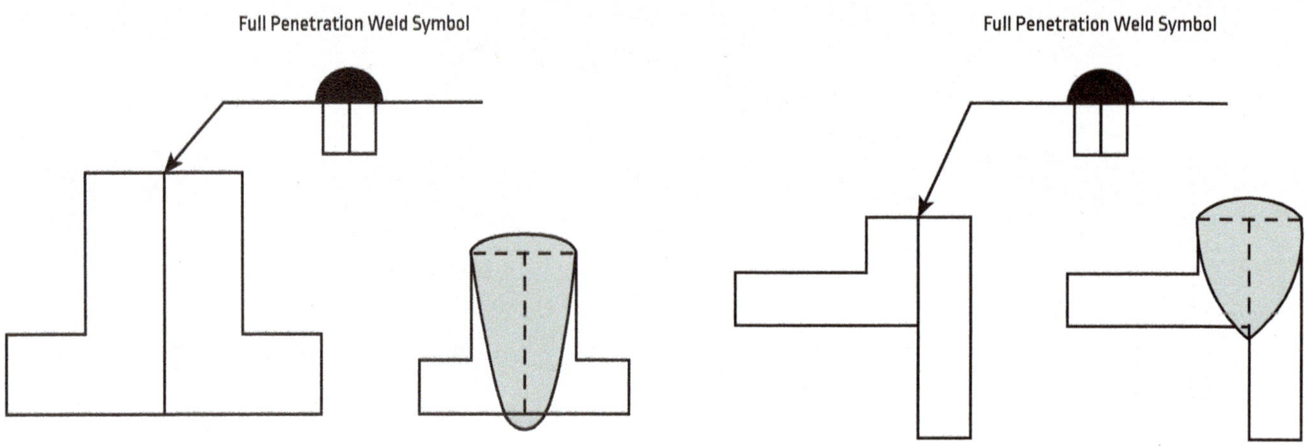

Figure 7.16 Full penetration edge welds fuse all members all the way through the joint.

Combination Edge Welds

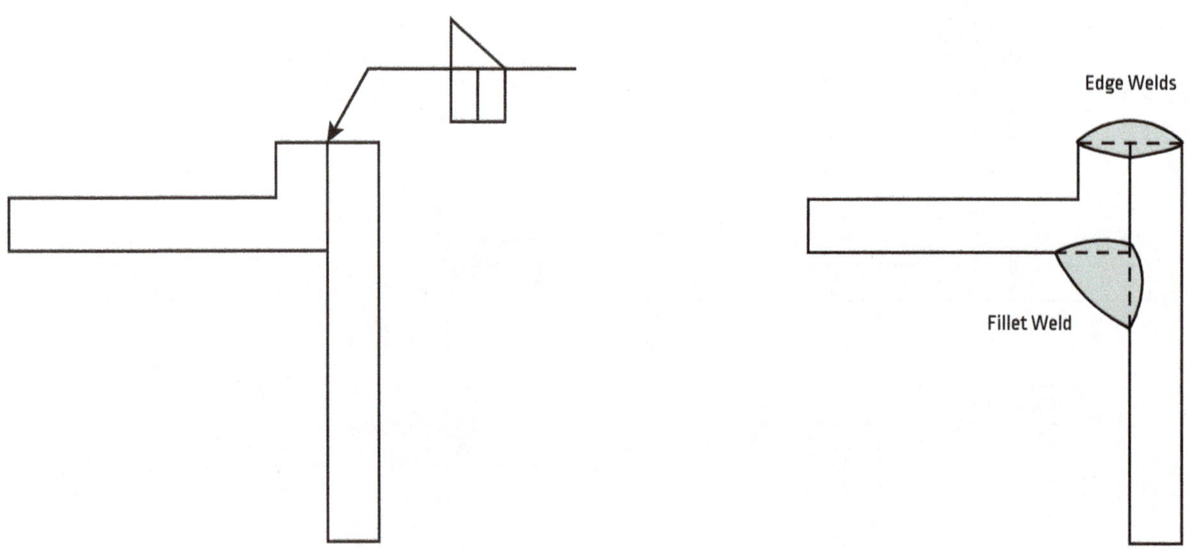

Figure 7.17 Combination welds can involve many different configurations. This depends mostly on the type of joint, material thickness and welding process.The symbol shows an edge weld on one side and a fillet weld on the other side.

Intermittent, chain intermittent and staggered intermittent edge welds are indicated the same as all other welds of this type. The welds are based on length and pitch.

Intermittent Edge Weld

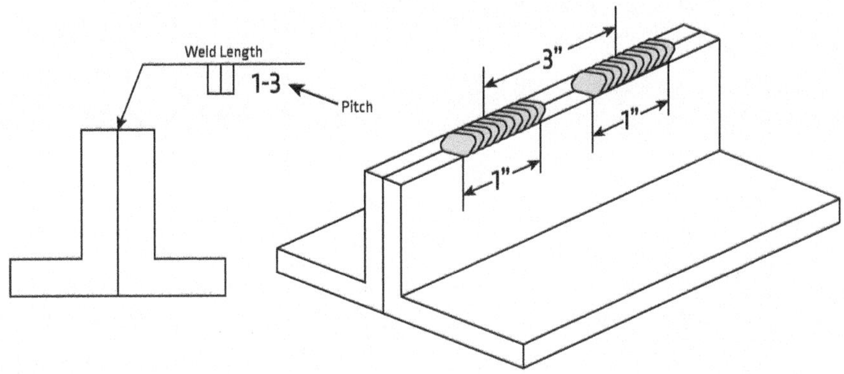

Figure 7.18 Intermittent edge weld 1 in. long, 3 in. center to center. Left, Weld symbol. Right, Sketch of weld.

Chain Intermittent Edge Weld

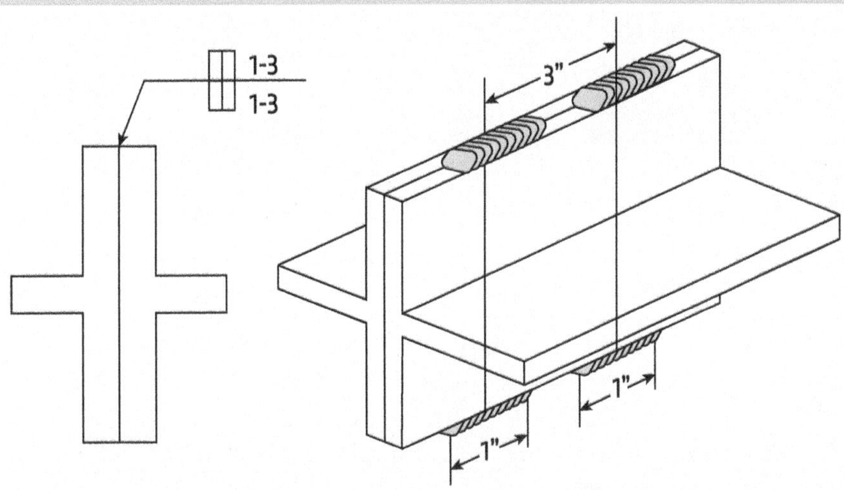

Figure 7.19 Intermittent edge weld 1 in. long, 3 in. center to center. Left, Weld symbol. Right, Sketch of weld.

Staggered Intermittent Edge Weld

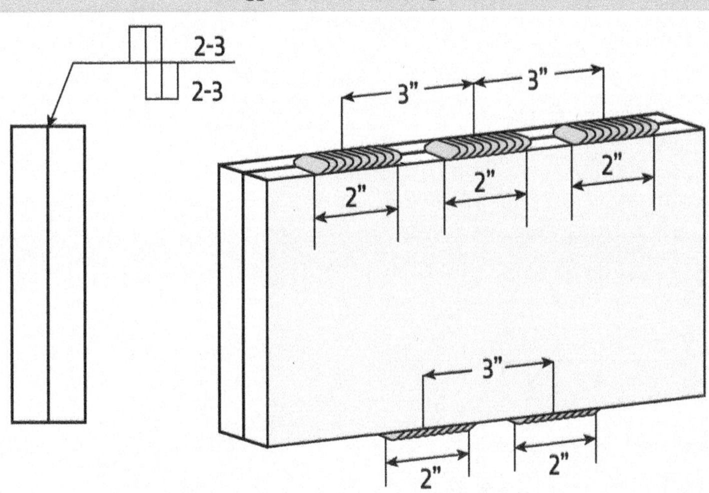

Figure 7.20 Staggered intermittent edge weld 2 in. long, 3 in. center to center. Left, Weld symbol. Right, sketch of weld.

Review Questions

Describe the weld symbol.

Example

1. 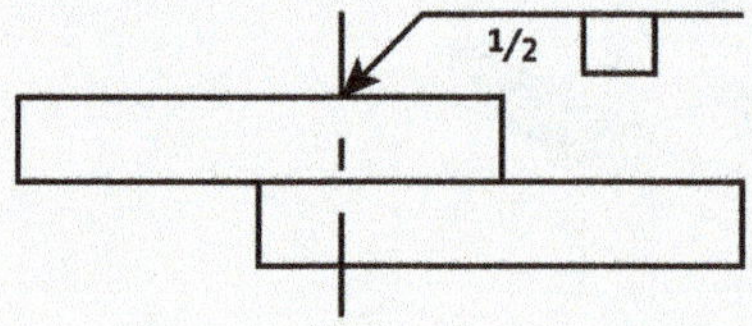

This is a 1/2" Plug Weld Arrow Side

2. 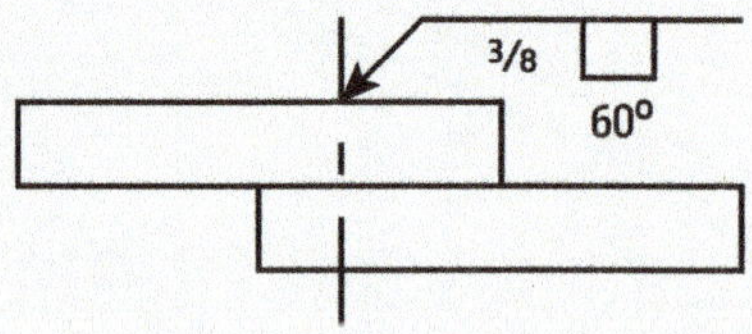

3. 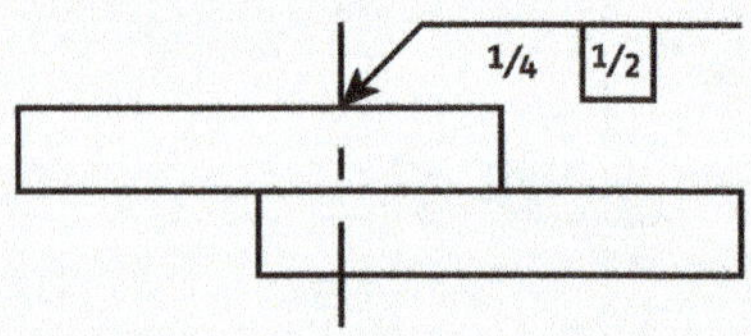

4. 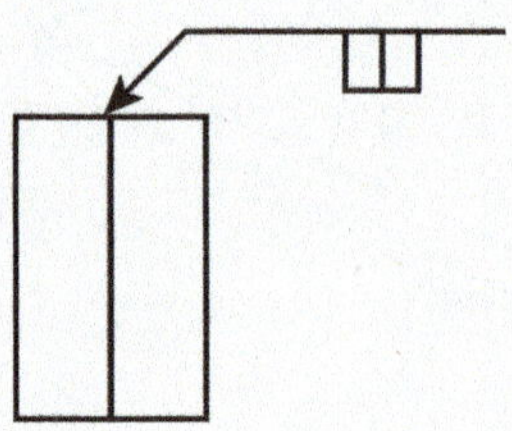

5. 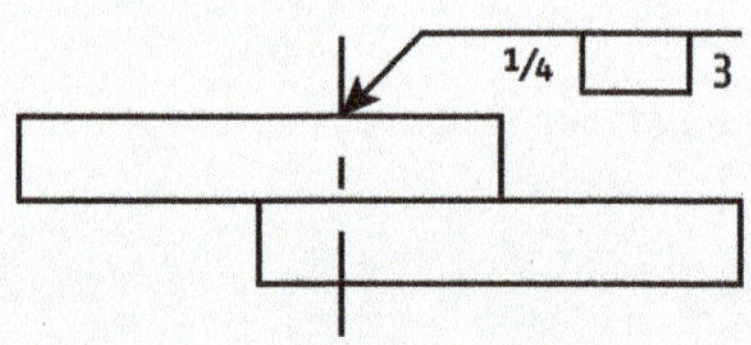

6. 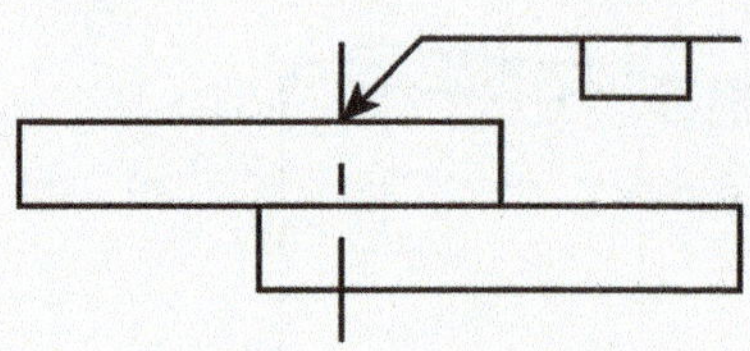

7. 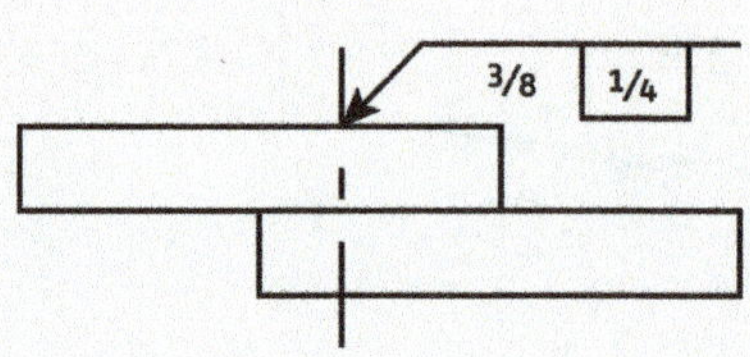

8. 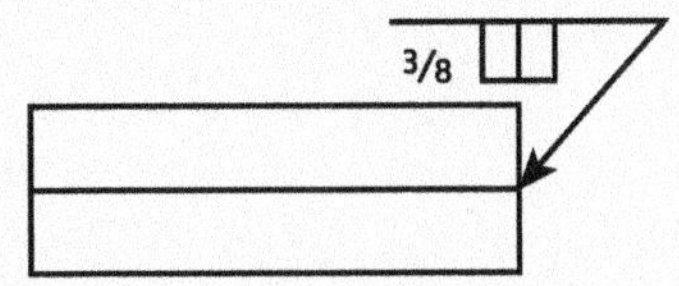

9. 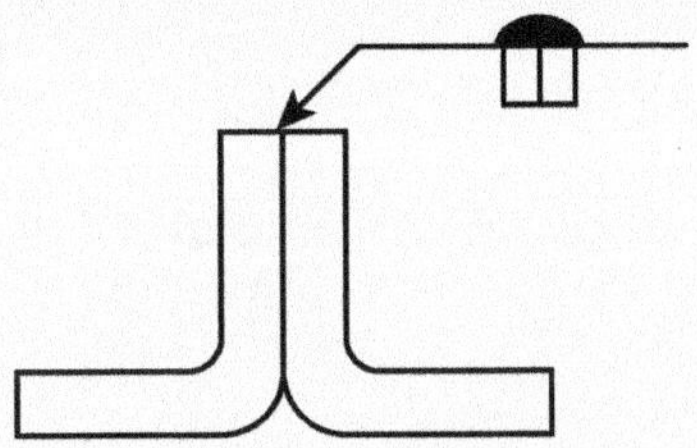

10. 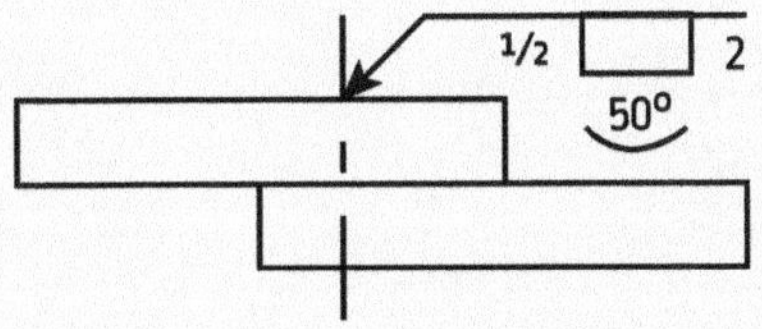

11. 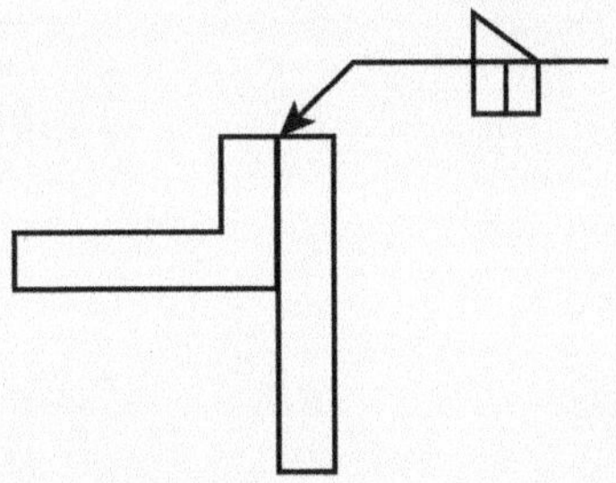

12. 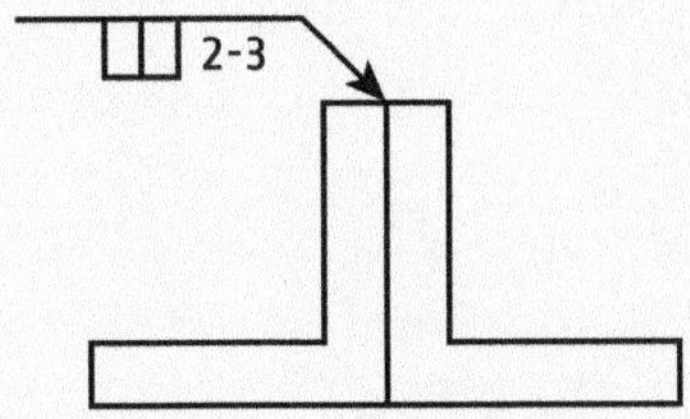

13. 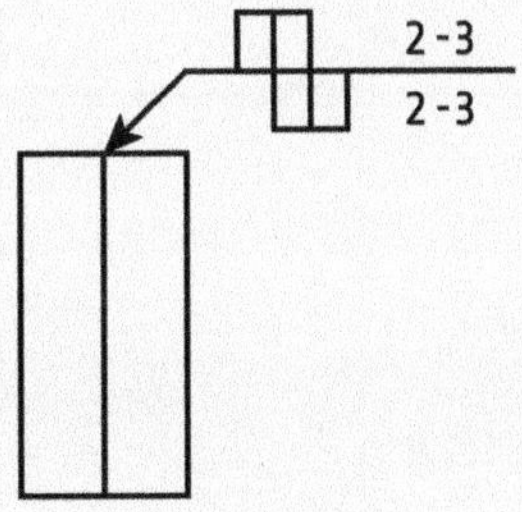

14. 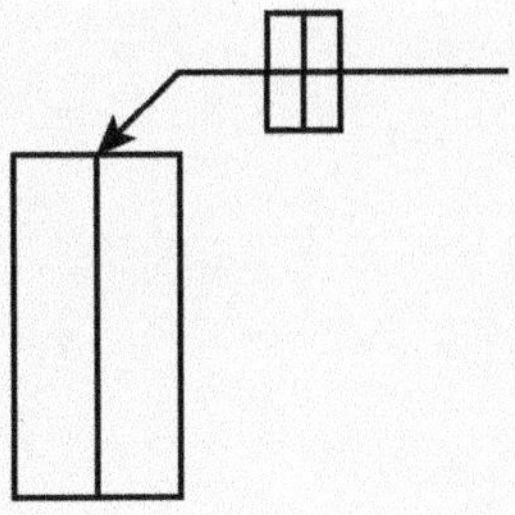

15.

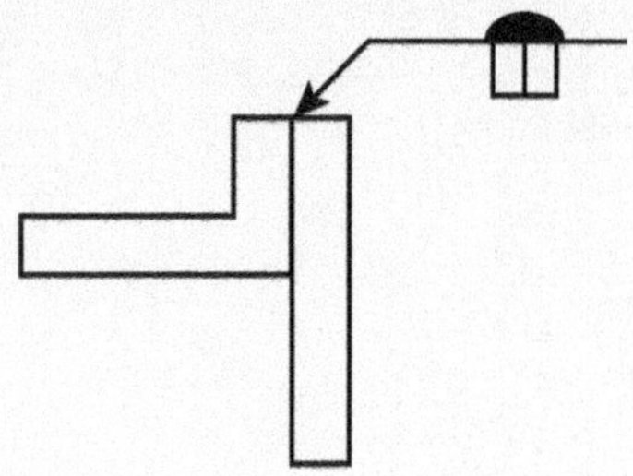

16.

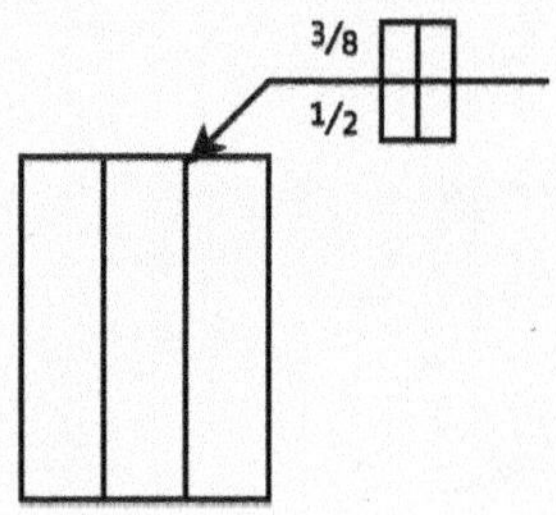

17.

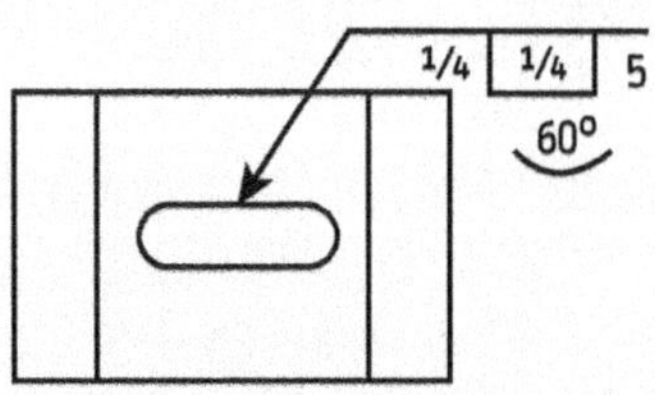

18.

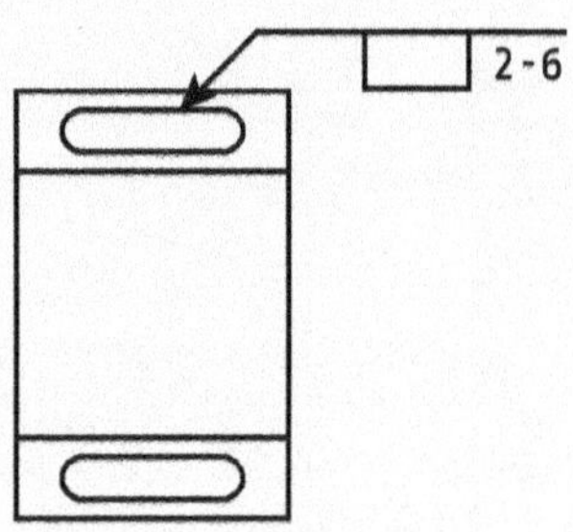

19.

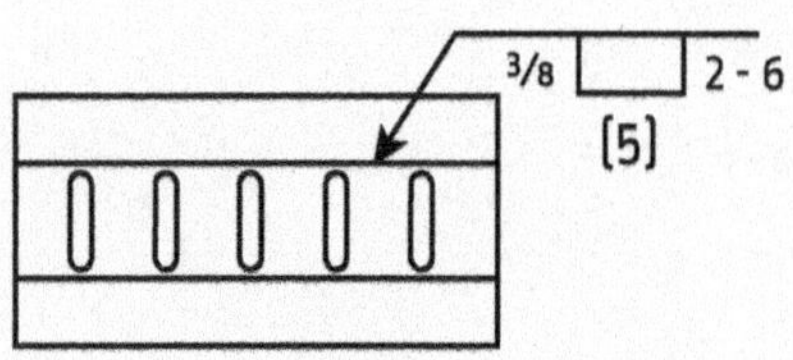

20.

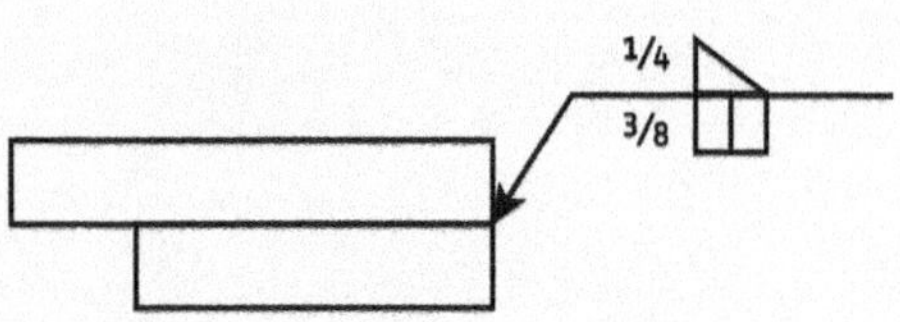

Notes

Notes

CHAPTER 8: Spot, Seam, Projection & Surfacing Welds

Chapter Objective

After learning the information in this chapter you will be able to:

- Identify spot, seam, projection and surfacing welds
- Identify the equipment used to perform these types of welds
- Determine location, spacing and size of welds
- Identify the number, size and strength requirements of the weld symbol

Key Terms

Spot welds
Seam welds
Projection welds
Surfacing welds
Stud welds
Hard facing

Resistance spot welds are formed by running electrical current through two pieces of metal forced together by two copper electrodes.

The resistance of the metal generates enough heat to create a weld nugget the size of the contact tips. There are two vital variables and current to making these welds: pressure and time. That is, how much pressure is used to hold the parts and for how long the weld is held.

Resistance spot welds are mostly used on sheet metal in the automotive manufacturing industry. A spot weld is shown on a drawing as a circle on either side of the reference line.

Spot Weld Shear Strength

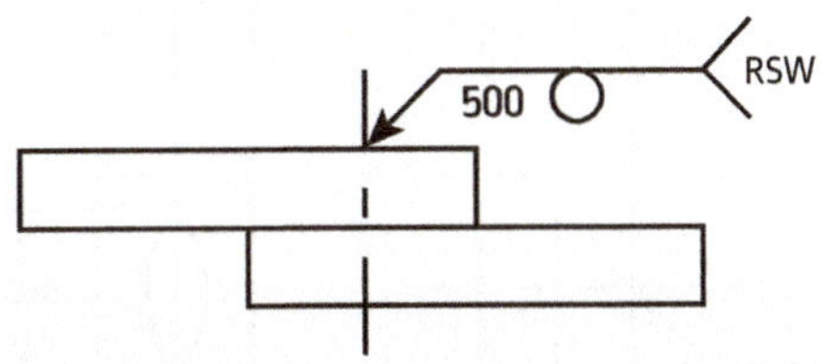

Figure 8.1 This symbol requires 500 lbs. of shear strength per spot weld minimum.

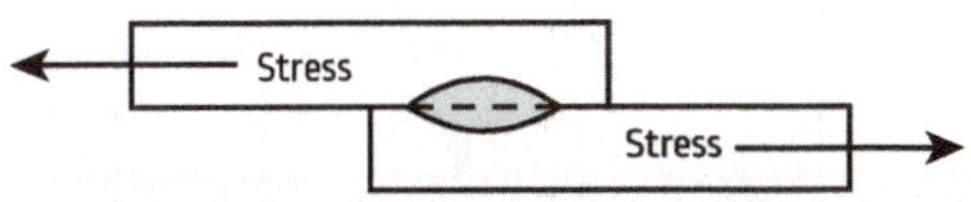

Figure 8.2 Shear strength means applying force in a way that tries to shear the weld off.

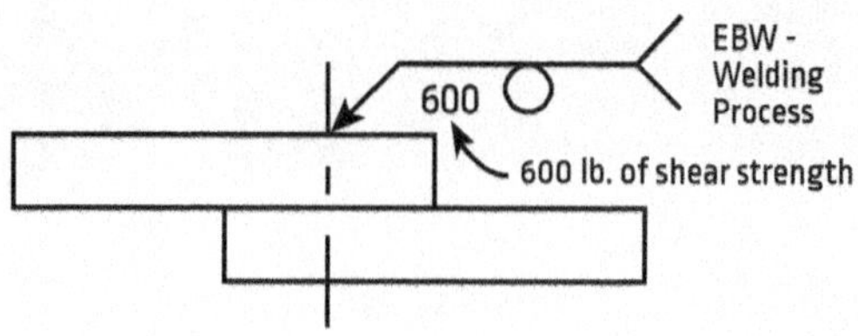

Figure 8.3 This symbol indicates an electron beam weld with a shear strength of 600 lbs. of shear load.

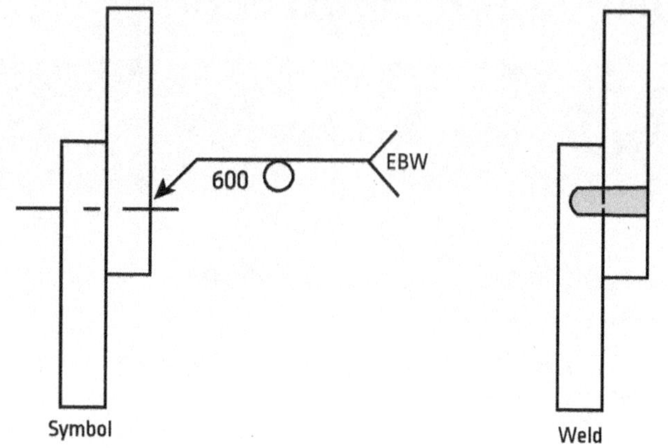

Figure 8.4 Symbol and sketch of weld for electron beam weld with a shear strength of 600 lbs. of shear load.

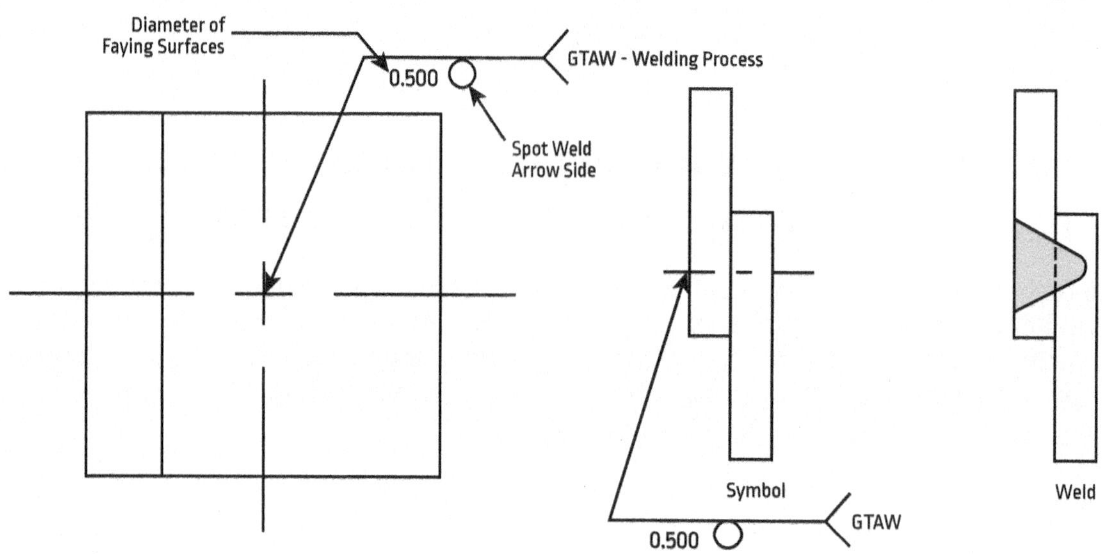

Figure 8.5 Spot welds can be used with various welding processes, which are indicated in the tail of the welding symbol.

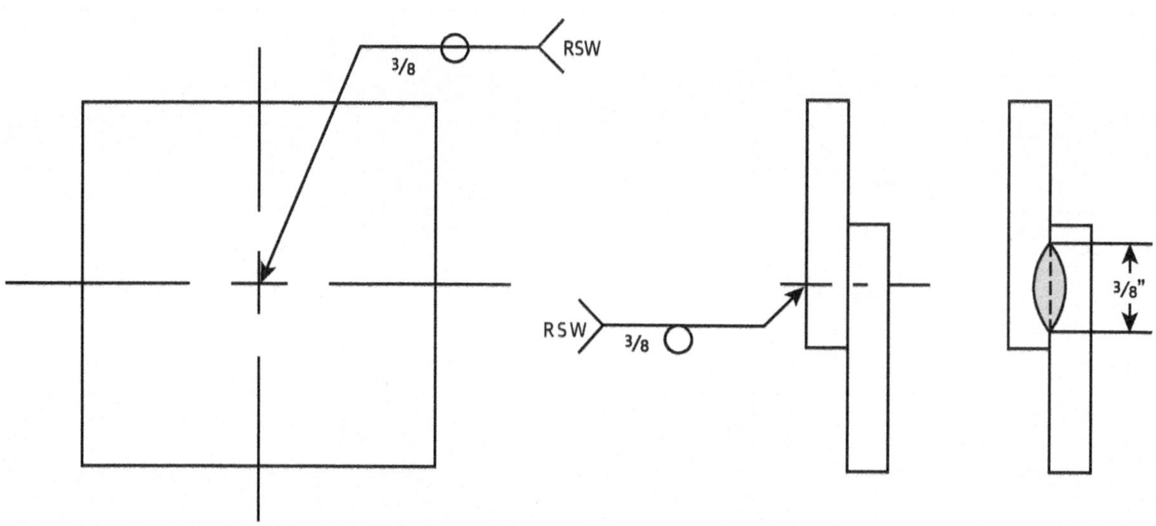

Figure 8.6 When the RSW symbol is placed on the line, there is no weld side preference. 3/8 in. spot weld. Left, Weld symbol. Right, Sketch of weld.

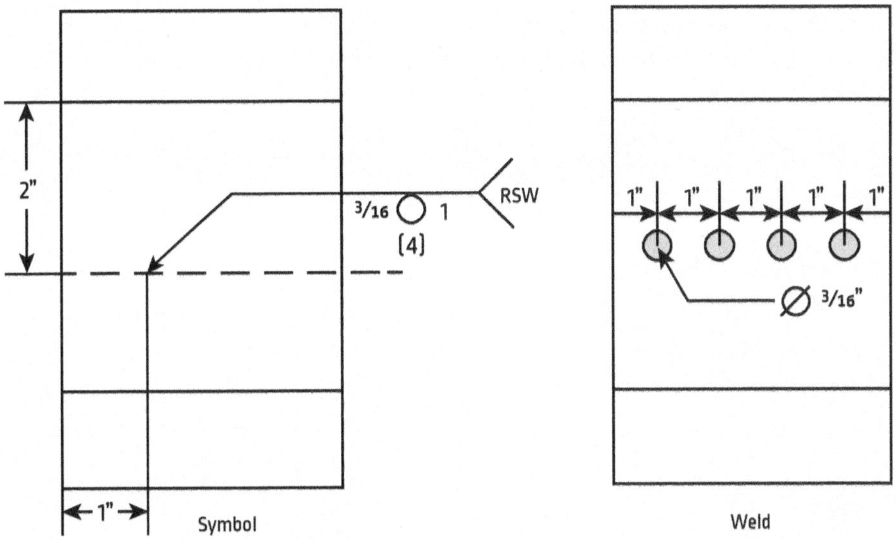

Figure 8.7 Series of four 3/16 in. spot welds with 1 in. pitch. Left, Symbol. Right, Sketch of weld.

The pitch of multiple spot welds is measured from center to center, the same as plug or slot welds. The pitch is always on the left of the weld symbol. The number of spot welds will also be indicated by a number in parentheses on the same side of the reference line as the weld symbol.

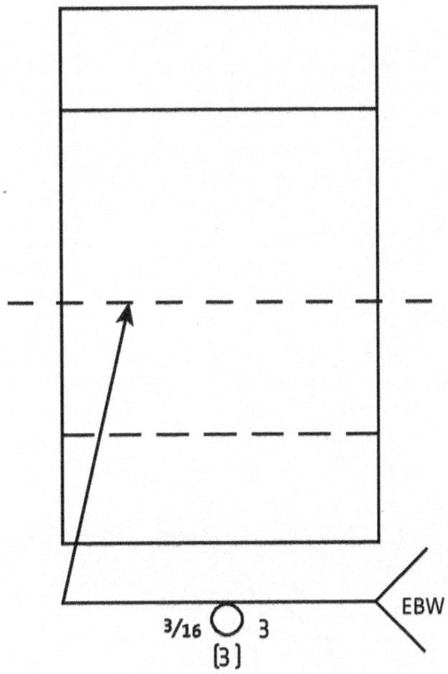

Figure 8.8 The number of spot welds in the series is indicated as a number inside parenthesis.

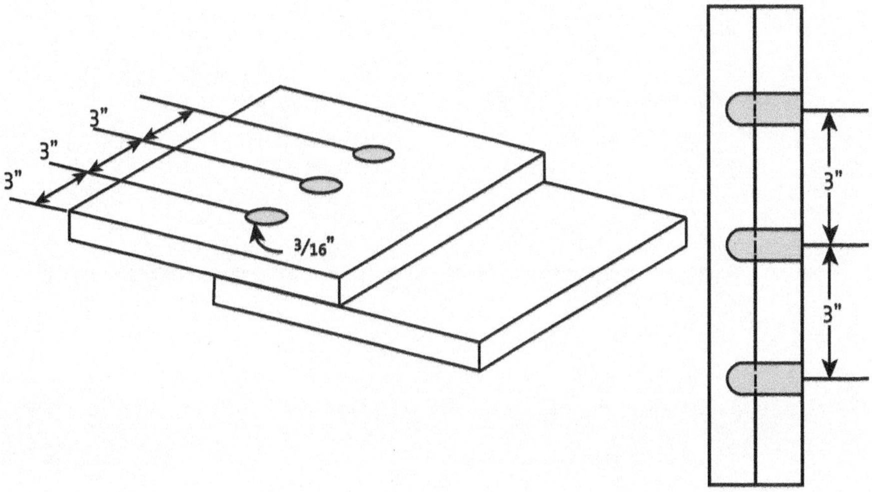

Figure 8.9 Sketch of weld whose symbol is shown in Figure 8.8. Series of three 3/16 in. spot welds with 3 in. pitch.

Seam Welds

Seam welds are similar to spot welds in that they can be formed using different welding processes but are usually formed using resistance welding. Seam welds are a series of overlapping spot welds made along a weld joint. A seam weld can also be one continuous weld. The weld symbol for a seam weld is a circle with two parallel lines running through it to resemble a seam. The welding process to be used is indicated in the tail of the welding symbol.

Seam Welds

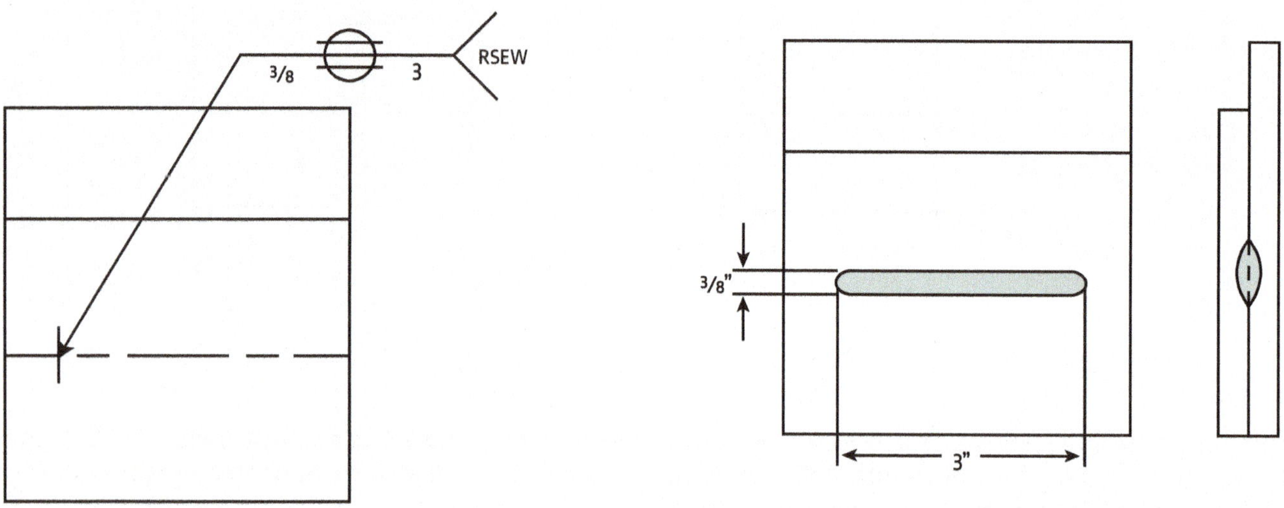

Figure 8.10 The symbol indicates a resistance seam weld 3/8 in. wide and 3 in. long. The process is indicated in the tail of the symbol.

	A		B		C
1.	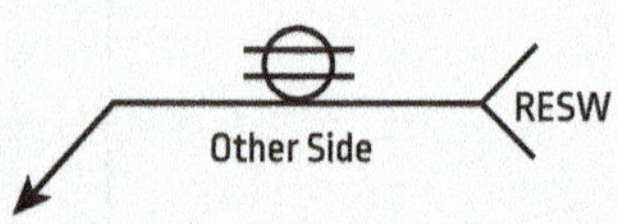	4.	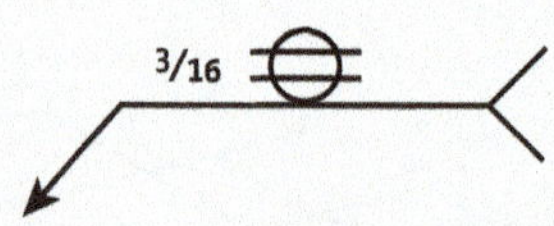	7.	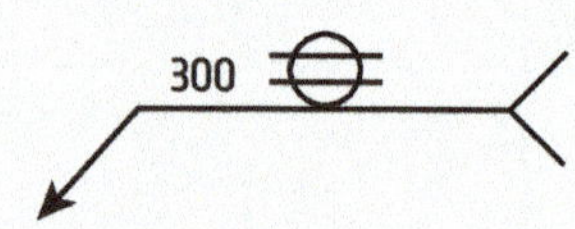
2.	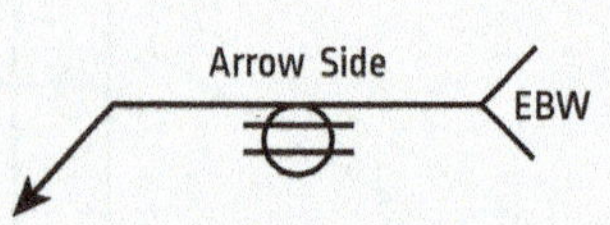	5.	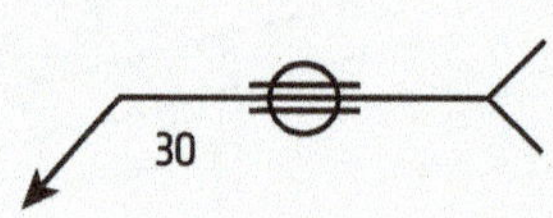	8.	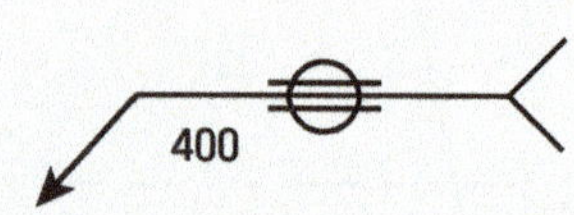
3.	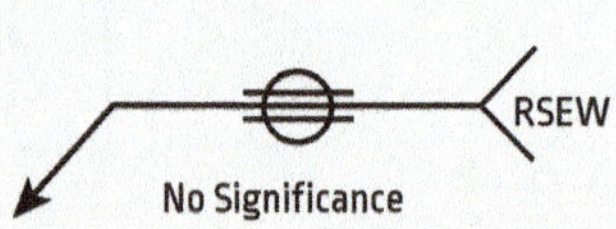	6.	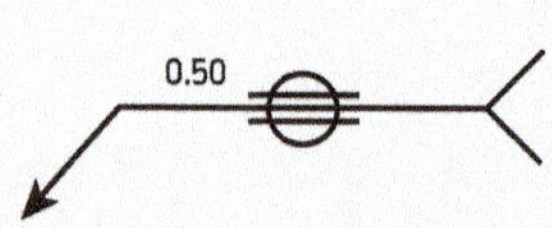	9.	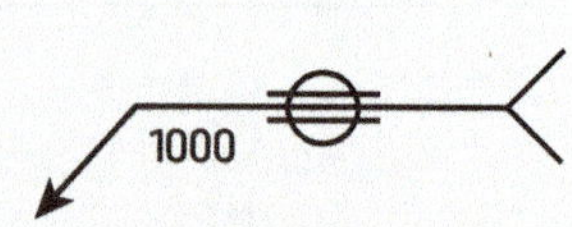
	The welding process and side of reference line are indicated on the symbol.		The size of the weld can be measured in fractions, millimeters or decimal forms.		The shear strength for seam welds is measured the same as the other resistance welds. It can be specified in pounds or Newtons.

Table 8.1

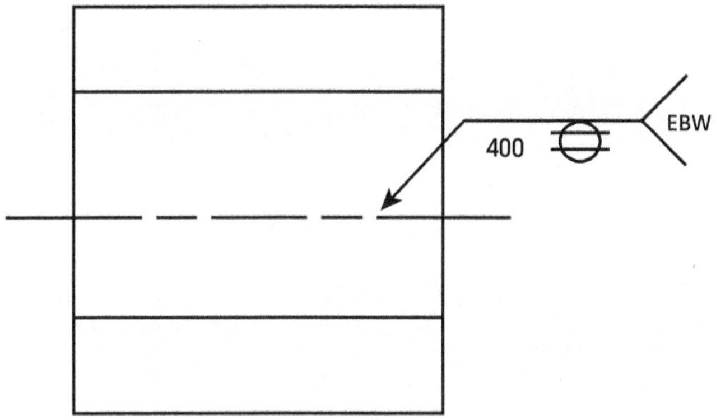

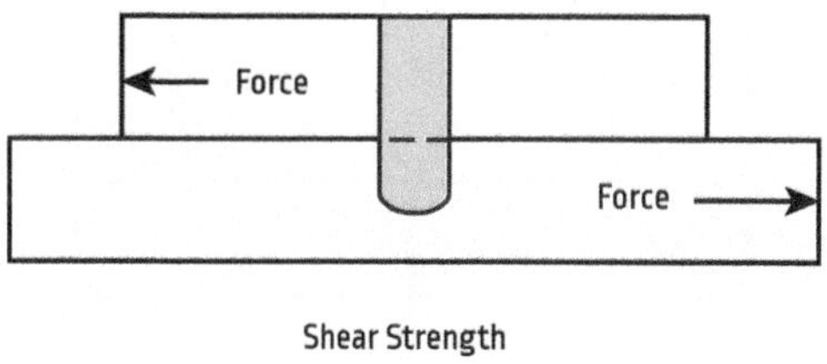

Figure 8.11 The number to the left of the symbol indicates 400 lbs. per linear inch of strength minimum is required for this weld.

Figure 8.12 Forces acting on the weld.

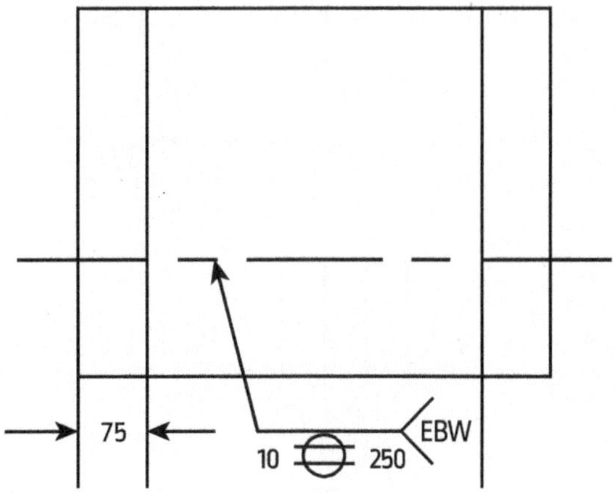

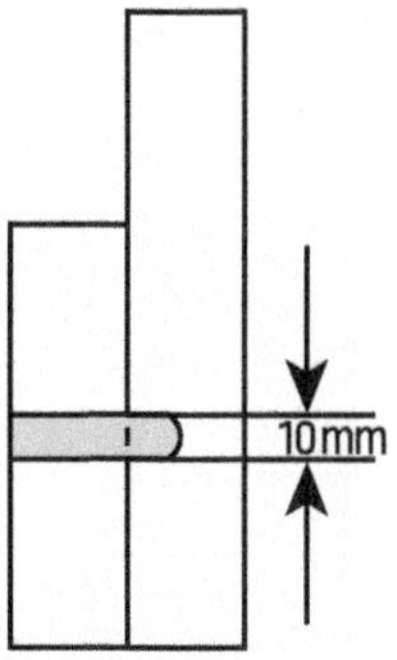

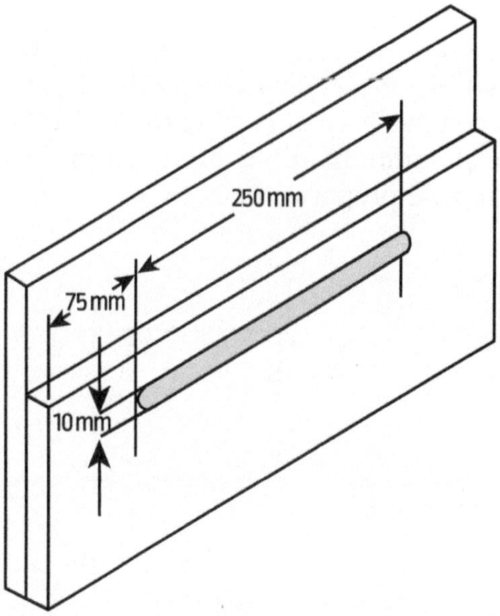

Figure 8.13 The drawing indicates measurements in millimeters. This measuring system is becoming used more and more often.

Figure 8.14 Cross section of seam weld shown in Figure 8.13.

Figure 8.15 Sketch of 10 mm seam weld, 250 mm long, as shown in Figures 8.13 and 8.14.

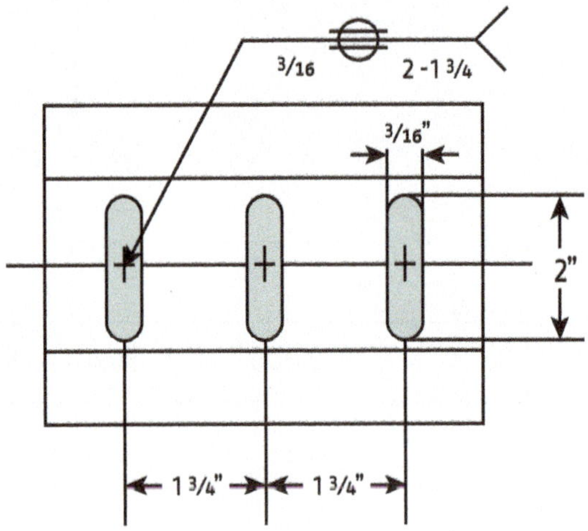

Figure 8.16 A series of 3/16 in. x 2 in. seam welds with 1-3/4 in. pitch.

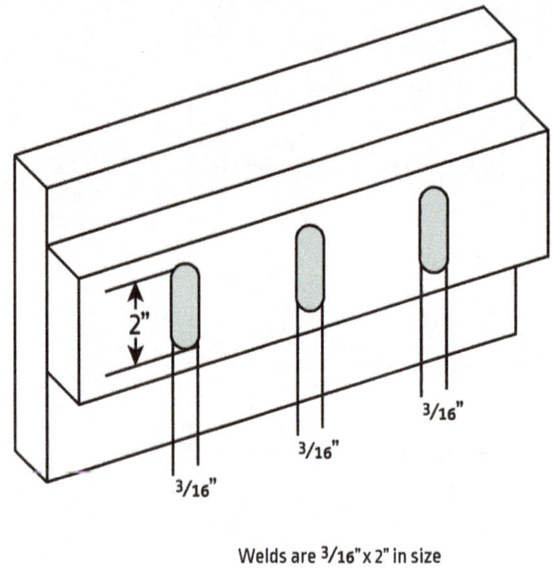

Figure 8.17 Sketch of weld whose symbol is shown in Figure 8.16.

Projection Welds

A projection weld is similar to a resistance spot weld, but instead of resistively heating to melt the metal which will fuse to form the nugget, a projection weld is formed using an electric arc and can be done with different welding processes. Projection welding uses a larger amount of current than does resistance spot welding. Projection welds can be used to weld threaded fasteners for bolting. This allows you to skip the expensive step of cutting or drilling holes. They are commonly used to weld fencing together. One of the most common applications for projection welds is stud welding, which is fast, accurate and requires little material preparation.

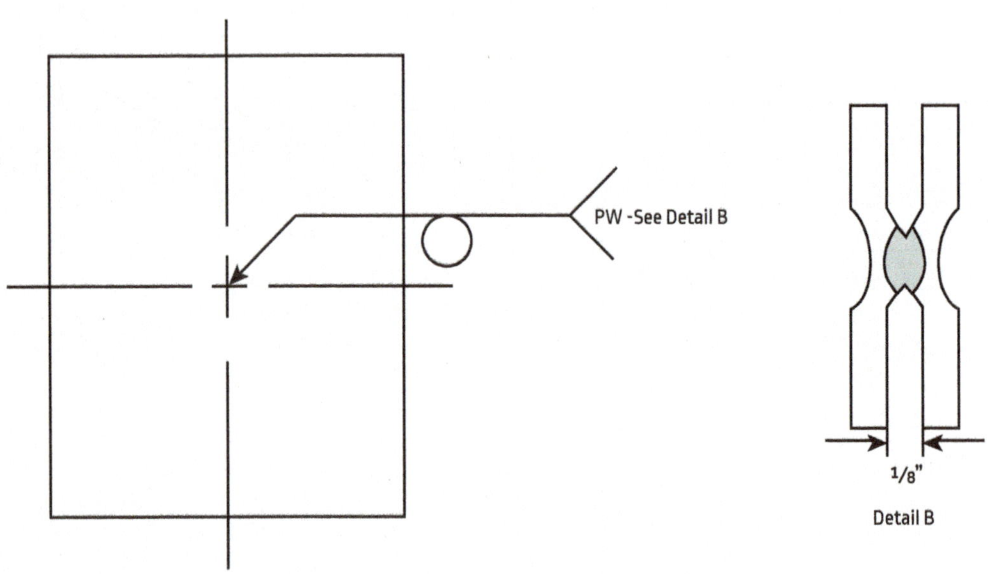

Figure 8.18 Projection weld. Left, Weld symbol. Right, Sketch of weld.

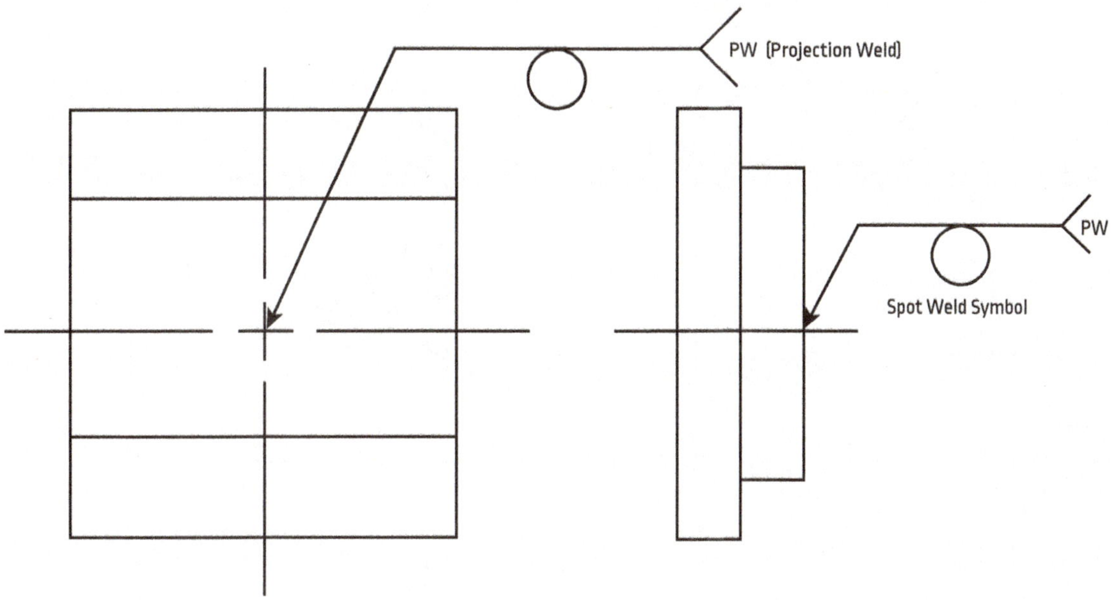

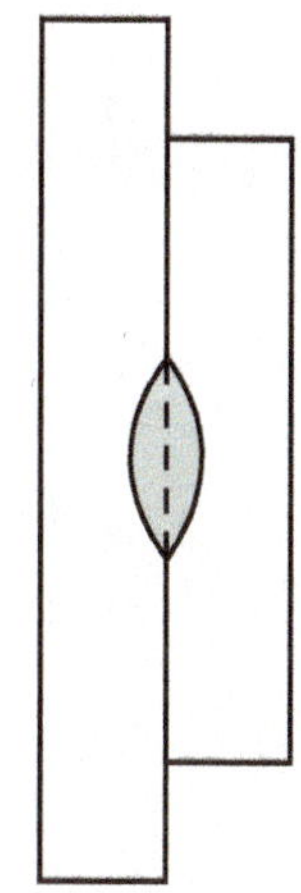

Figure 8.19 The symbols for PW and RSW are similar. The difference is indicated in the tail.

Figure 8.20

The first time I ever saw projection welding, it was being used to pull dents out of sheet metal on a car. This projection weld gave the repair technicians the ability to remove dents even though they had no access to the back of the panel.

Stud Welds

One of the most common applications for projection welds is stud welding. Stud welds are shown on a welding symbol as a round circle with an X through the center. Stud welds are common in the construction and manufacturing industry. The automotive industry uses small hand held stud guns to weld round pull-tabs for removing dents from body panels.

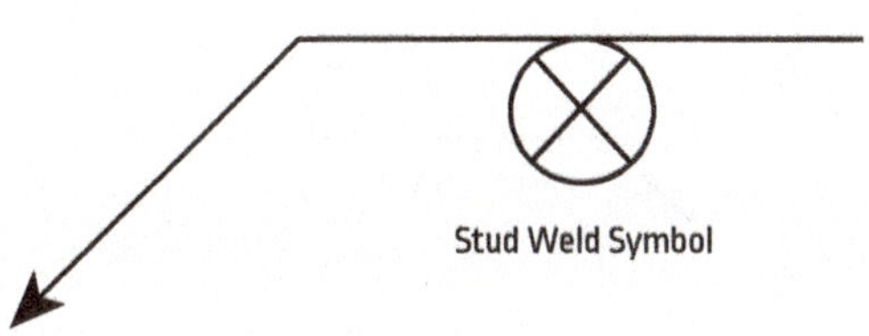

Figure 8.21 Stud welds are always specified from the arrow side.

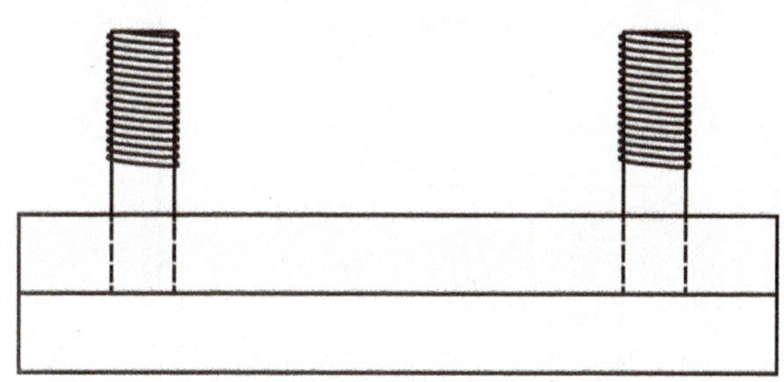

Figure 8.22 One application for stud welding is to weld bolts or screws to plate.

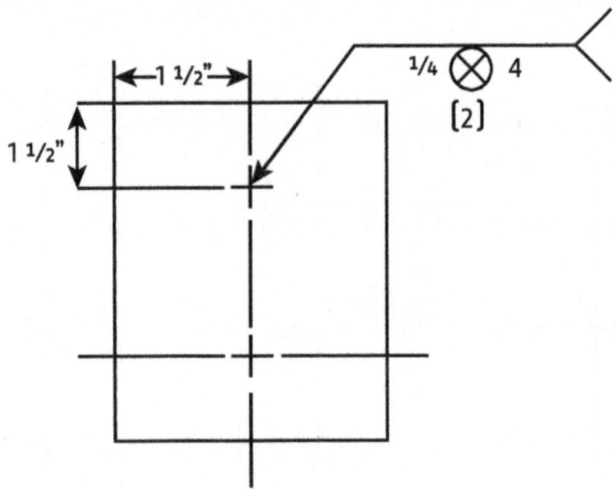

Figure 8.23 The symbol indicates there are two 1/4 in. studs on four in. centers

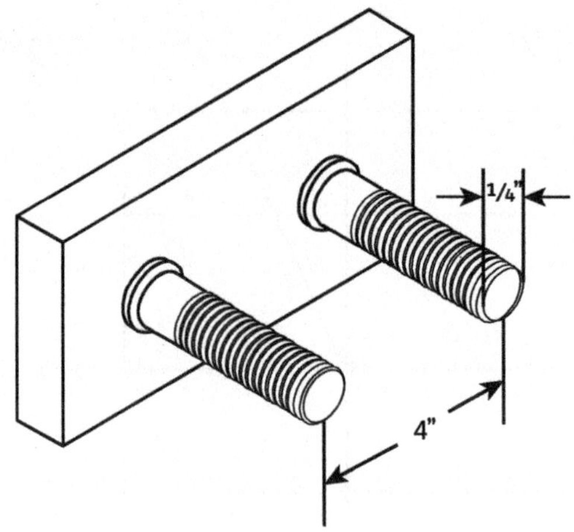

Figure 8.24 Sketch of stud weld shown in Figure 8.23.

A stud weld is indicated on a welding symbol and the size of the stud weld is indicated on the left side of the weld symbol.

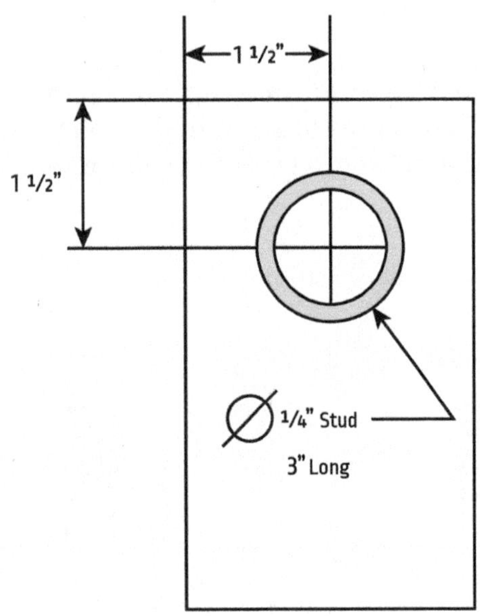

Figure 8.25 The placement of information is similar to many of the different weld types we have seen throughout the book. The length of the stud is usually indicated by a note or supplemental drawing.

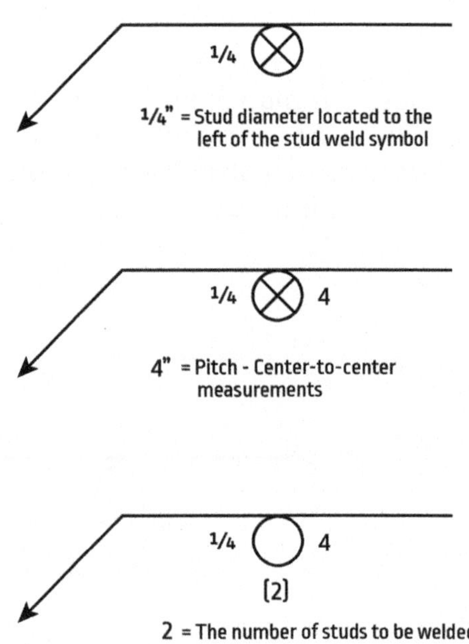

Figure 8.26 How to read a stud weld symbol.

Many times multiple studs are welded to the same piece in very specific intervals. These welds are similar to other intermittent welds except we do not have a weld length, only pitch. The welding symbol will show the stud size to the left of the symbol and the "pitch" is shown on the right side of the weld symbol. Pitch is measured on a center-to-center basis. If the weld symbol shows a 2 in. pitch, then the center of each stud is 2 in. apart.

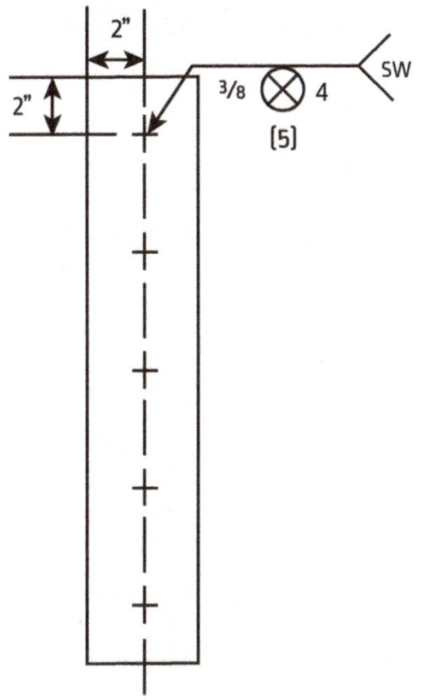

Figure 8.27 Illustrations showing multiple stud welds and the symbol showing how many, pitch, and stud size.

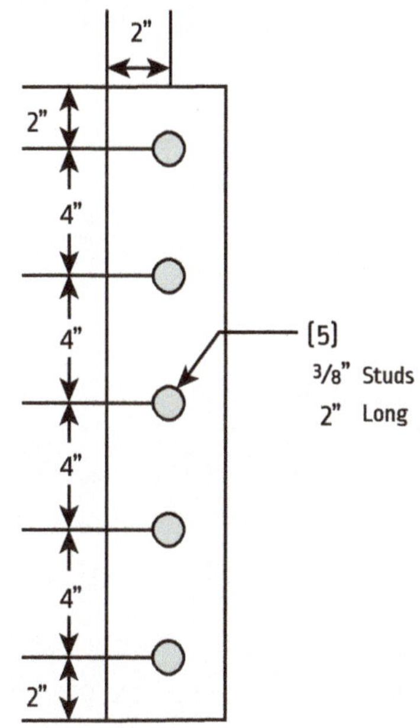

Figure 8.28

Surfacing Welds

Surfacing welds are also known as weld beads or stringer beads. Stringer beads are run consistently and continuously with approximately a 50% overlap. Surfacing welds are often used for building up surfaces for machining or in heavy wear areas such as the cutting edges of heavy equipment blades and buckets. Surfacing welds are shown on a drawing as two open half circles and the weld bead size will always be to the left of the welding symbol.

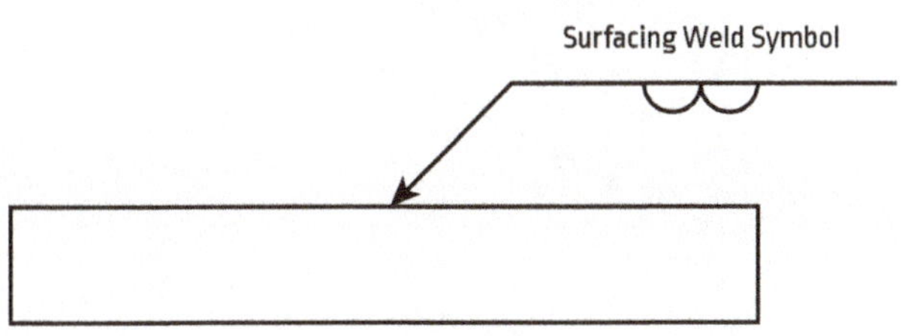

Figure 8.29 Surfacing welds are measured in built-up thickness of the weld layer. If there is no size of surface welded area indicated it is assumed the entire surface is to receive surfacing welds.

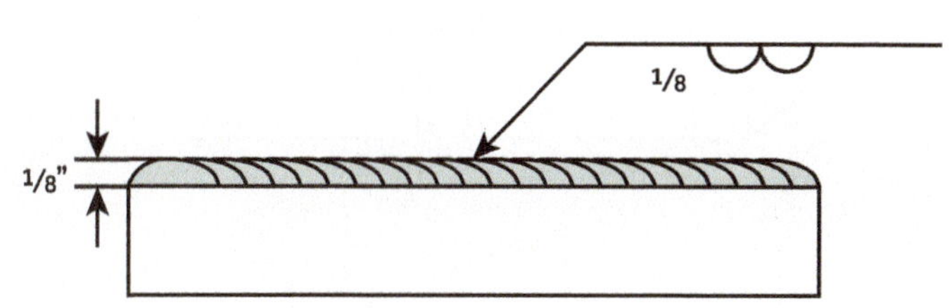

Figure 8.30 Sketch of weld whose symbol is shown in Figure 8.29.

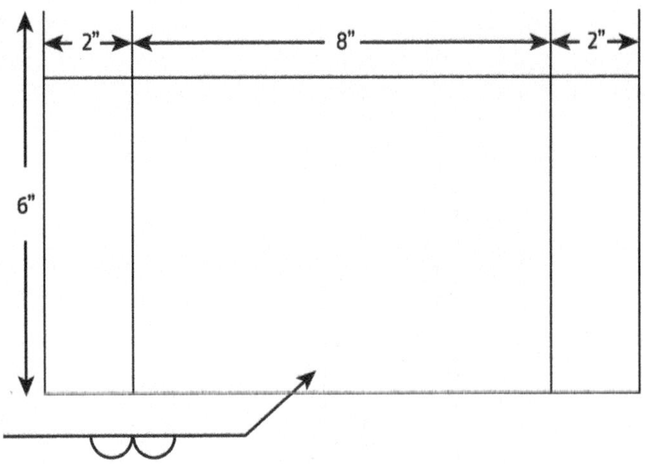

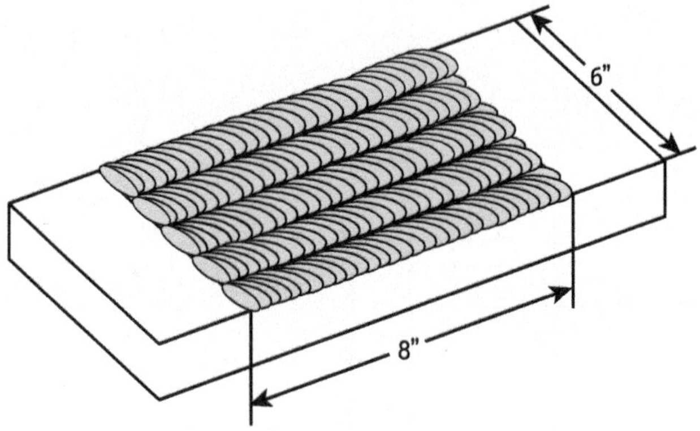

Figure 8.31 Drawing indicates only 6 in. x 8 in. of the area receives weld. Weld symbol indicates surface weld.

Figure 8.32 Sketch of weld indicated in Figure 8.31.

Buttering is a process where a surfacing weld is used to tighten tolerances on a worn out part, increase the dimensions of a part or prepare edges for dissimilar welds. Buttering uses compatible weld metals to build edges.

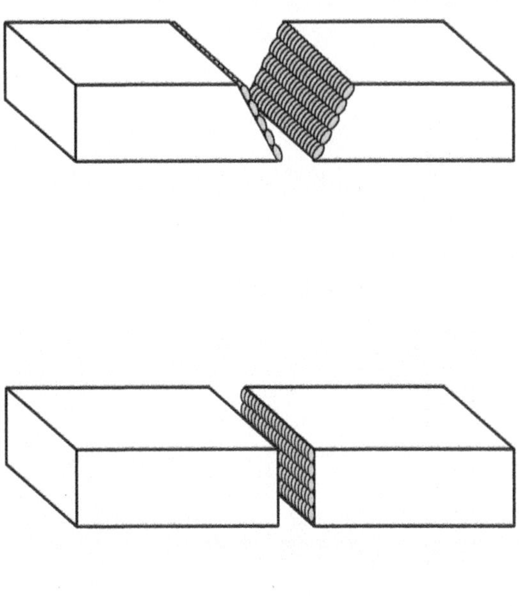

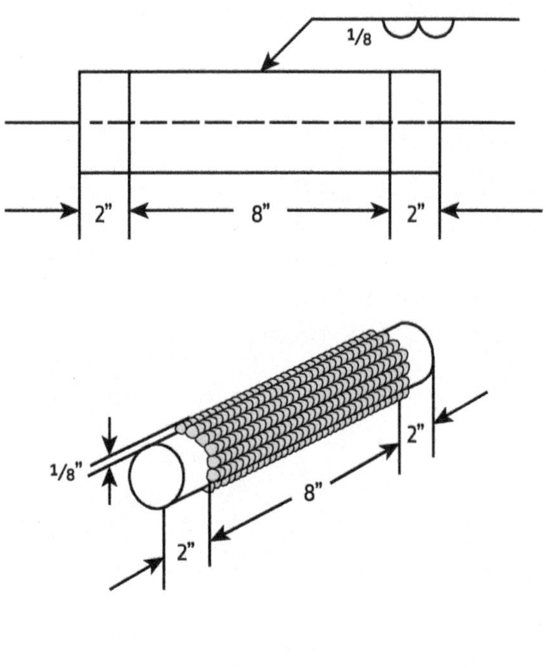

Figure 8.33 Buttering edges can close tolerances and make the parts usable again instead of a scrap bin donation.

Figure 8.34 Surfacing welds are used to reduce wear and build edges but can also be used to build parts that can be re-machined into usable components again.

Review Questions

Describe the weld symbol.

Example

1. 

Spot weld with 400 lb shear strength on Arrow Side

2. 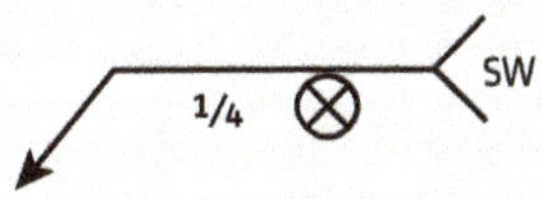

3. 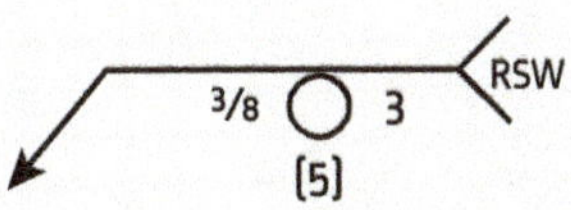

4. 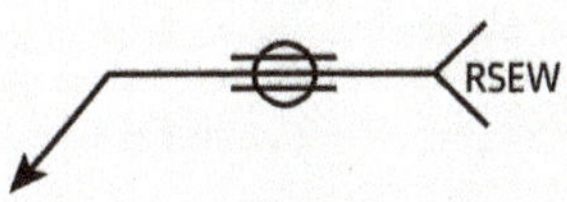

5. 

6. 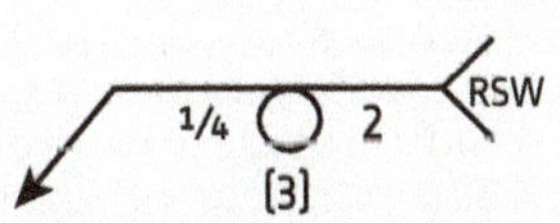

7. 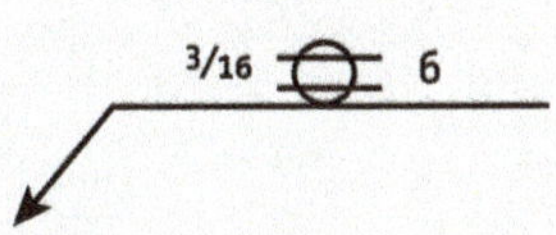

8. 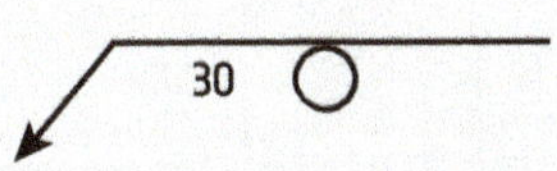

9. 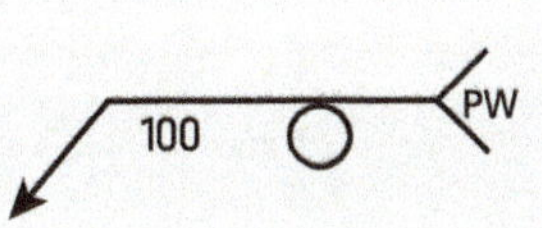

10. 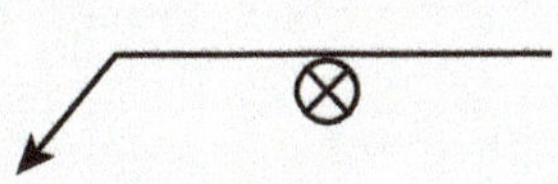

11. 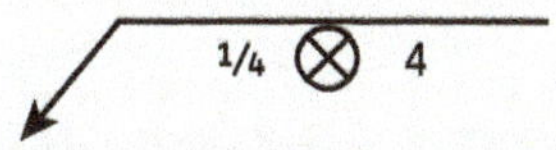

12. 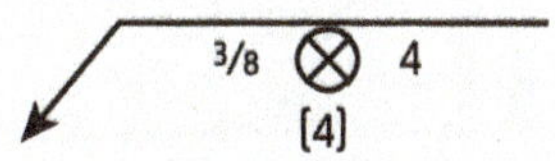

13. 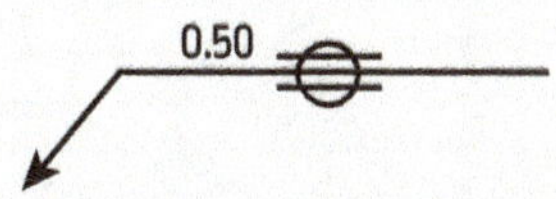

14. 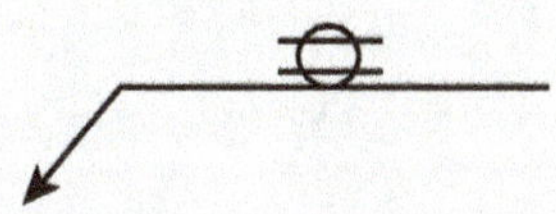

15. 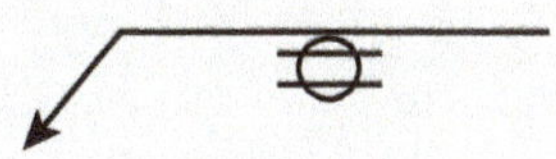

16. 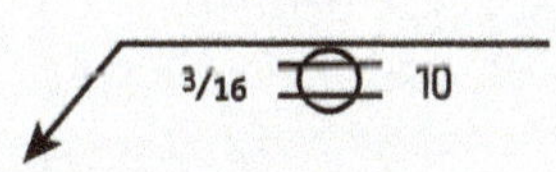

17. 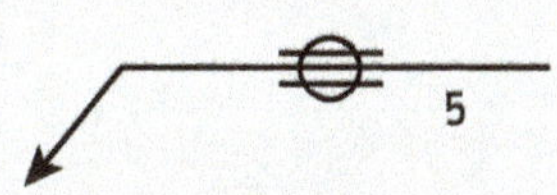

18. 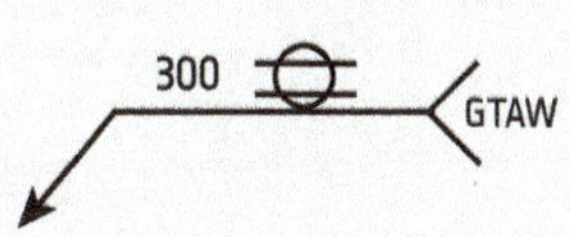

19. 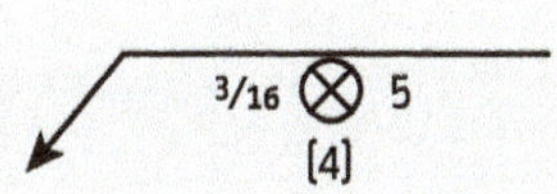

20. 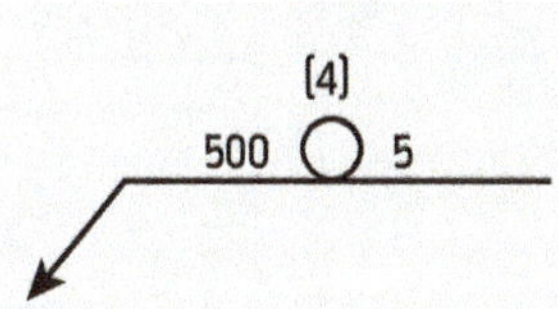

Review Questions

Identify each part of the weld symbol.

Example

1.

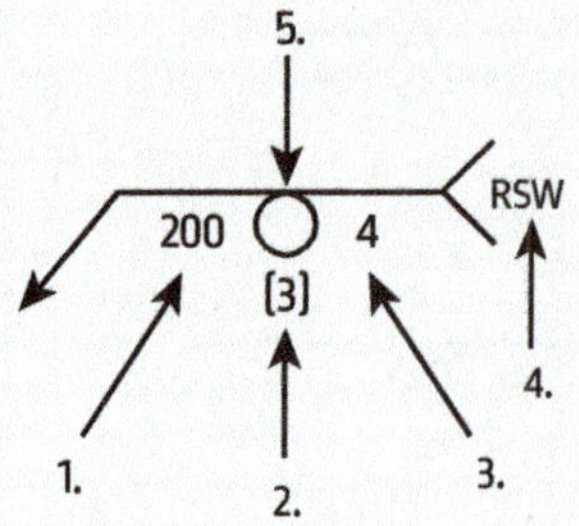

1. 200 lb/in Shear Strength
2. 3 Spot Welds
3. 4" Centers
4. Resistance Spot Weld
5. Spot Weld Arrow Side

2.

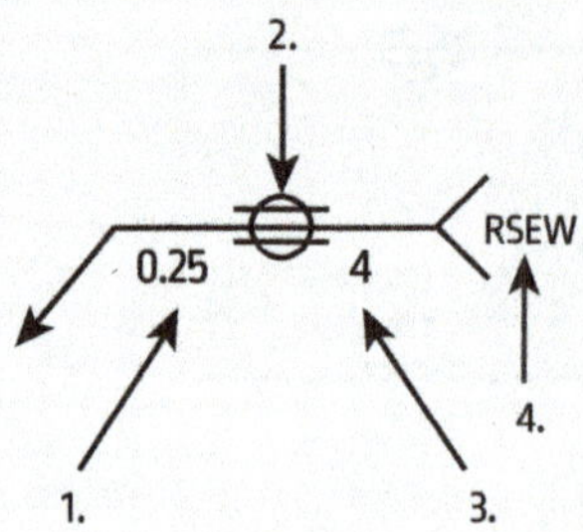

1. ____________________
2. ____________________
3. ____________________
4. ____________________

3.

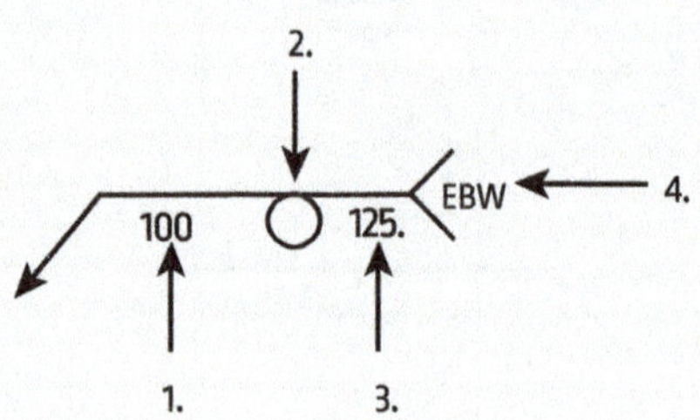

1. ____________________
2. ____________________
3. ____________________
4. ____________________

4.

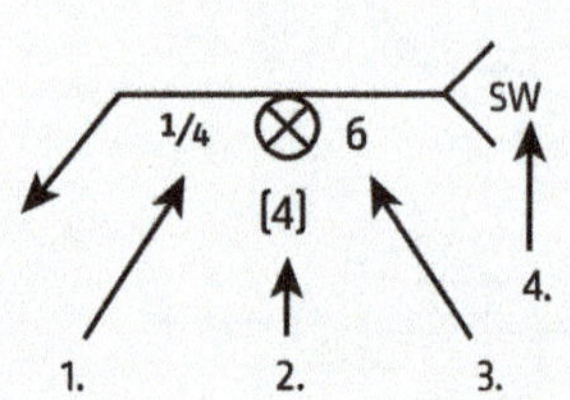

1. ____________________
2. ____________________
3. ____________________
4. ____________________

5.

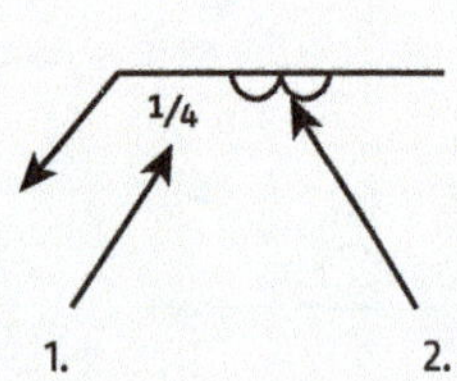

1. ____________________
2. ____________________

Notes

Notes

CHAPTER 9: Industry Materials

Chapter Objective

After learning the information in this chapter you will be able to:

- Identify different material types and classifications
- Compare the different structural steel shapes
- Understand the different weldable metals
- Identify the governing organizations on material specifications
- Identify different alloys and their advantages and disadvantages

Key Terms

Base metal
Alloys
Structural steel
I-beam
C-channel
Angle
Tubing
Pipe
Bar stock
Sheet steel

One of the biggest components of the Industrial Revolution was the mass production of steel. Although steel was being produced long before the Industrial Revolution, it was not until this time that huge steel plants were constructed and steel makers began making thousands and thousands of tons of steel for railroads, bridges, tools, equipment and one of the largest consumers, the construction industry.

The welding and fabrication industry uses many different types of materials. As a welder, a shop supervisor or even a material purchaser, we need to know how to speak intelligently about the materials we need. In industry, we use differently shaped materials in a wide variety of ways, so we must be specific when ordering the materials. It is important that you know the difference between pipe and tube or plate steel and sheet steel. This chapter will not attempt to cover every type of material available. We will focus more on steel and the different shapes of steel commonly used in the fabrication industry.

One of the biggest components of the Industrial Revolution was the mass production of steel. Although steel was being produced long before the Industrial Revolution, it was not until this time that huge steel plants were constructed and steel makers began making thousands and thousands of tons for railroads, bridges, tools, equipment and one of the largest consumers, the construction industry.

Structural Steel

Structural steel is used in all kinds of construction and building erection. Steel manufacturers make steel with very close tolerances and constantly monitor the chemical compositions of the molten steel. The American Iron and Steel Institute (AISI) and the Society of Automotive Engineers (SAE) have developed a classification numbering system to identify the different types of carbon and alloy steel. Steel manufacturers shape these steels and alloys into different shapes for different uses and different industries. Most metal is made into certain shapes by forming, rolling, drawing through dies, stamping, forging, extruding, spinning or machining. Manufacturers may use a variety of terms to refer to the shape of their materials. Below is a discussion of the most commonly used structural steel shapes.

Figure 9.1 Structural steel in bridge construction.

I-Beams

I-beams are used to provide strength and support over long spans in building construction, bridges and many other applications. I-beams are categorized in several ways. The designation for a standard I-beam is the letter S. Some beams have a much wider flange than the standard I-beam. These beams are designated with the letter W. The letter M identifies different miscellaneously-shaped beams. The letters HP identify beams characterized by weight per foot and nominal dimensions over the flanges. An example would be a W 15 x 25. This is indicating that the beam size over the flanges is 15 in. and the beam weighs 25 lb. per foot.

I-Beams	
Structural Shape	Symbol
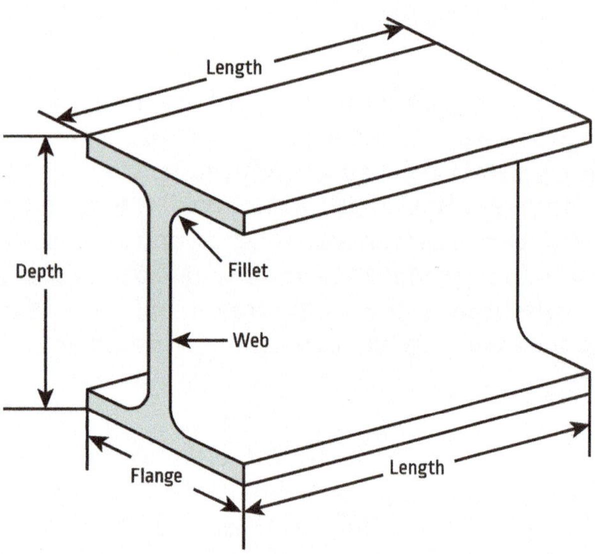	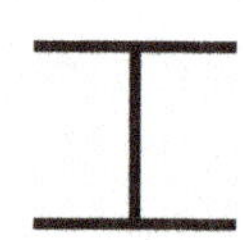

Size Specification				
Depth in Inches		Weight per foot - lbs or kilograms		Length in Feet
15 "	x	40	x	10'

Figure 9.2 An I-beam with dimensions.

Flange Beams	
Structural Shape	**Symbol**

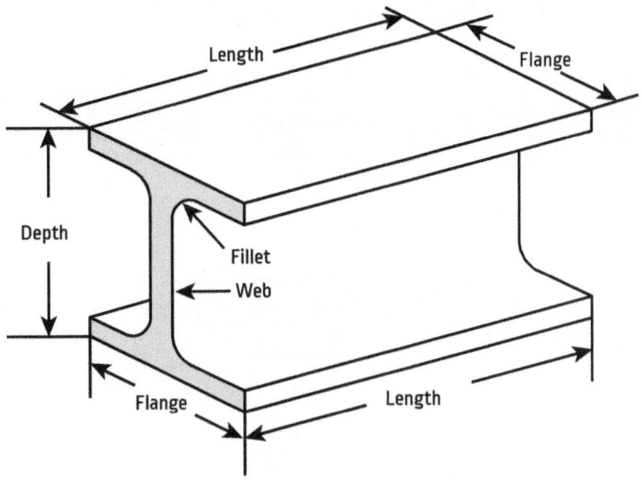

WF or M

Size Specification		
Depth in Inches	Weight per foot - lbs or kilograms	Length in Feet
30" x	85 x	28'

Figure 9.2 (Continued) An I-beam with dimensions.

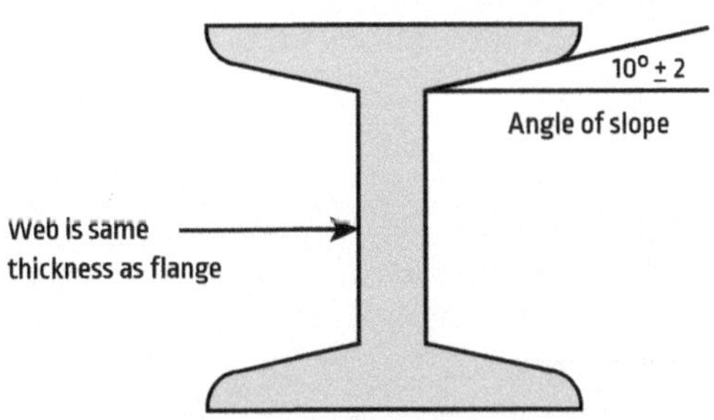

(S) = Standard Beam

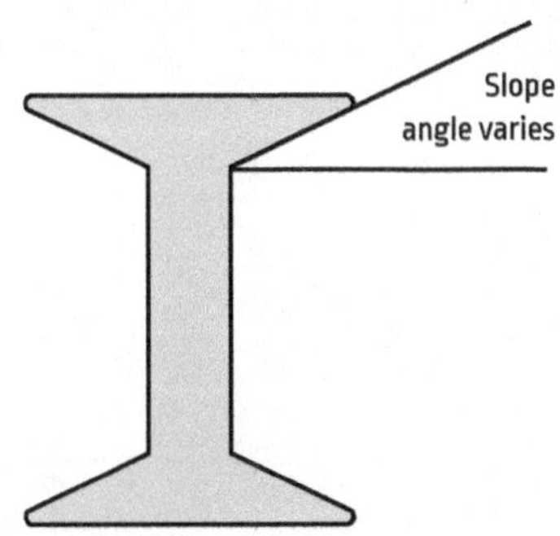

(M) = Miscellaneous beams other than HP, wide flange or standard beams

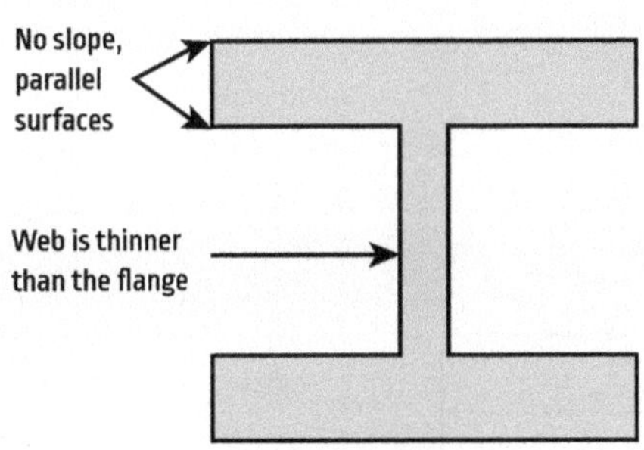

(W) = Wide Flange Beam

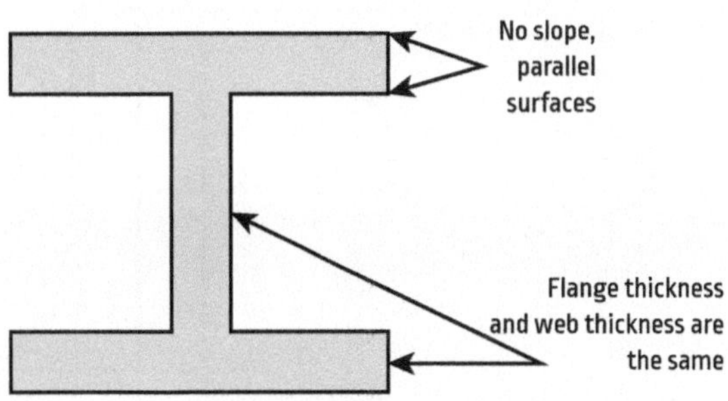

(HP) Bearing Pile Beam

Figure 9.2 (Continued) An I-beam with dimensions.

C-Channel

C-channel is another structural steel that is used to help strengthen or support other types of materials, such as beams. The American designation for C-Channel is a letter C. The letters MC designate different miscellaneous types of channels. Depending on the application, two pieces of C-channel can be welded back to back to create a version of an I-beam. We measure channel from the heel to the toe of the flange.

C-Channel	
Structural Shape	Symbol

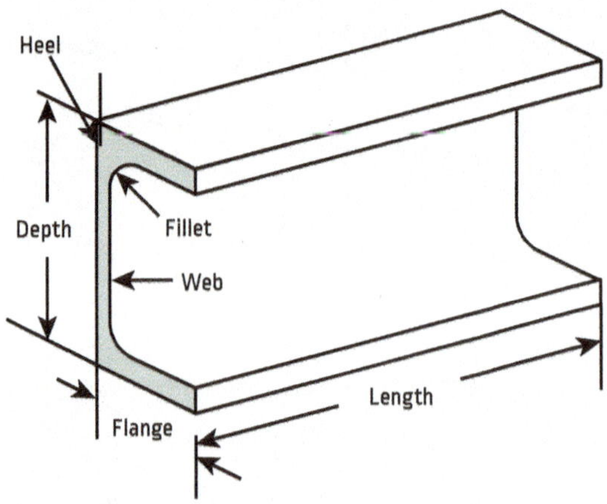

Size Specification			
Depth in Inches	Flange in Inches	Weight per ft. in lbs.	Length in Feet
10" x	5" x	20 x	20'

Figure 9.3 C-Channel with dimensions

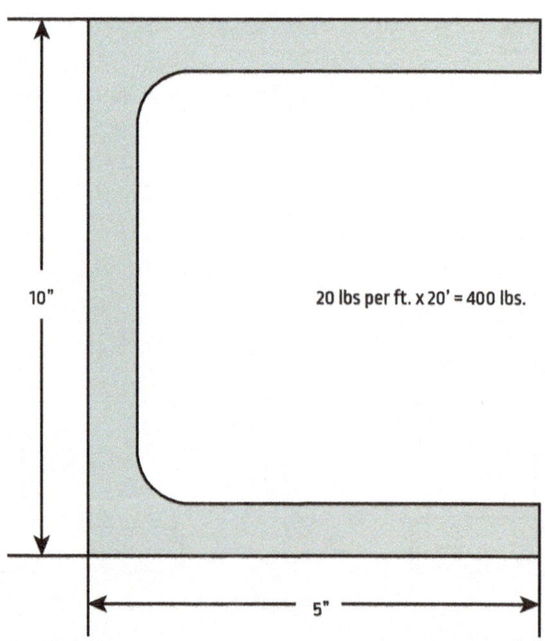

Figure 9.3 C-Channel with dimensions

Angle Stock (L Shapes)

Angle stock is structural steel shaped like an L. Angle stock's legs can be equal or unequal. Angle stock is a strong shape that provides strength when stressed in different directions. Ninety-degree angle plays a big role in structural steel construction. Angle stock is often used to strengthen and reinforce larger structural steel components. It provides a great deal of rigidity when a 90° angle is required. Angle stock can be cut and re-welded, bent, folded and notched. It is a versatile and strong material that can be used without an extensive amount of labor. Angle stock is measured from the heel to the end of the legs, and the leg measurements are always followed by the thickness of material. For example, angle 3 x 2 x 1/8 would indicate angle stock with one 3.0 in. leg and one 2.0 in. leg and made out of 1/8 in. material. If angle is 3 x 1/8, this would indicate angle stock with two 3.0 in. legs, made out of 1/8 in. material.

L - Beam

Structural Shape

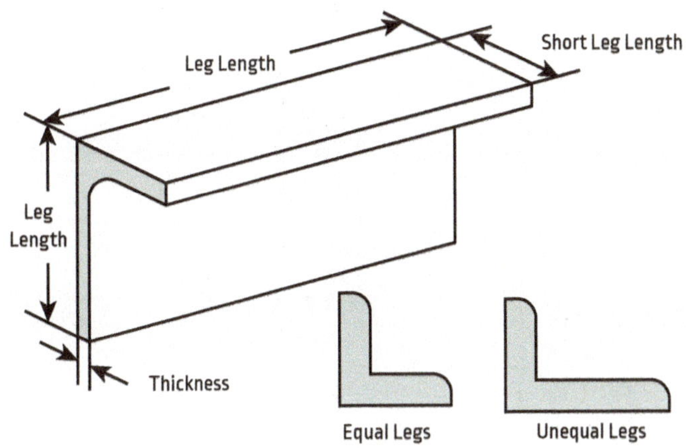

Symbol

Size Specification						
Leg Width in Inches		Leg Width in Inches		Thickness in Inches		Length in Feet
4 1/4"	x	4 1/4"	x	1/4"	x	20' 4"

Figure 9.4 Angle and measurements.

Tee

T-shaped structural steel materials are essentially shaped like a letter T. The T is measured by the nominal depth and flange size. These shapes provide more strength than angle at a lower cost than an I-beam. T-shaped materials are usually cut from I-beams or machined by the supplier.

T - Beam

Structural Shape

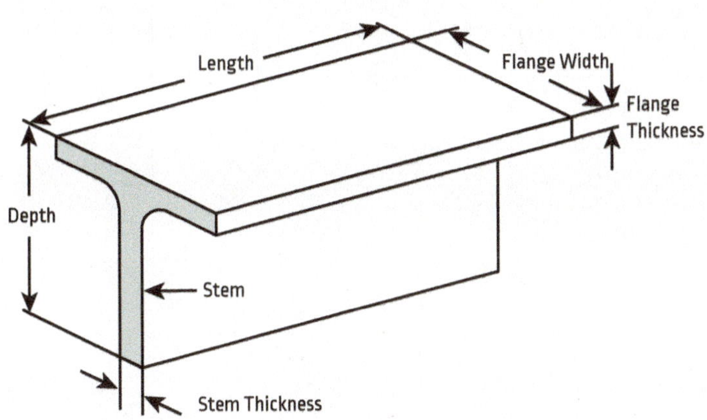

Symbol

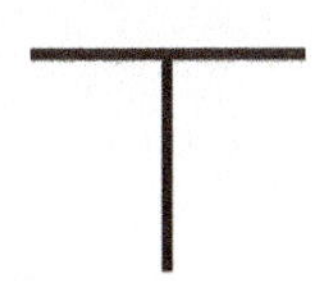

Size Specification						
Width of Flange in Inches		Depth of Stem in Inches		Thickness of Flange in Inches		Length in Feet
3"	x	2"	x	3/8"	x	18'6"

Figure 9.5 T-shape beam and measurements.

Tubing

Tubing comes in different shapes including square, round and rectangular and provides strength for reinforcement and structural applications. Measurements for square and rectangular tubing are given as length x width x wall thickness. The wall thickness and size of the tubing is directly related to its strength. An example would be a square tube 2 in. x 2 in. x 3/16 in. This would be a tube 2.0 in. in width, 2.0 in. in depth and with a 3/16 in. wall thickness.

Square Tube

Structural Shape

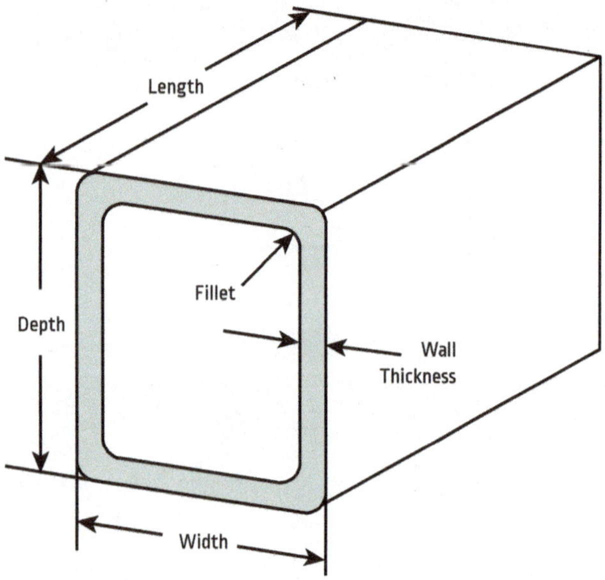

Symbol

Tube

Size Specification

Width in Inches		Depth in Inches		Wall Thickness in Inches		Length in Feet
8"	x	8"	x	1/4"	x	18' 10"

Figure 9.6 Square tubing and measurement.

Rectangle Tube

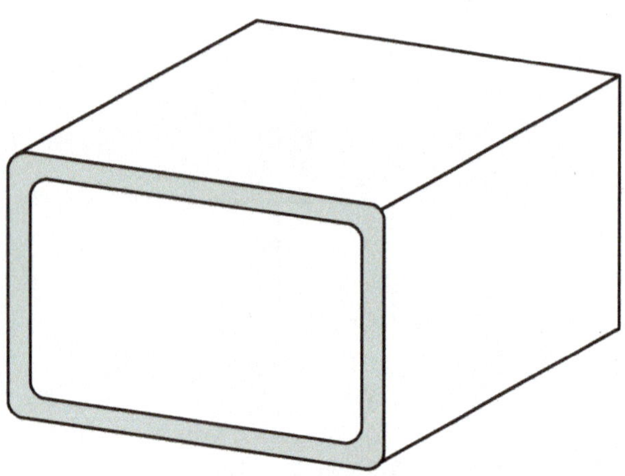

Round Tube

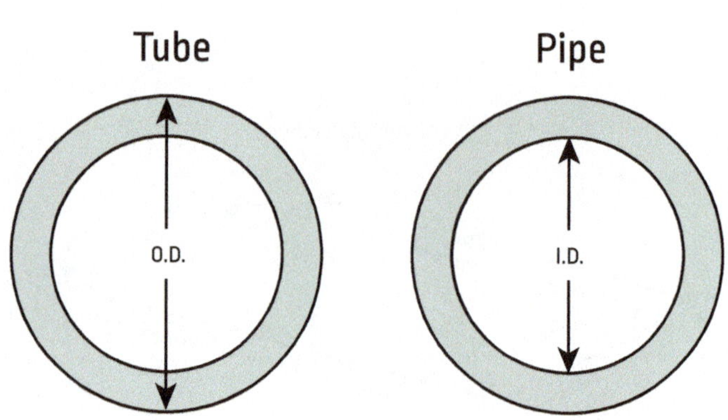

Figure 9.7 Rectangular tubing is measured the same as square tubing.

Round tubing sizes are based on the wall thickness and outside diameter (OD) of the tube. Pipe, on the other hand, is measured using the inside diameter (ID). Tubing is strong and is used in many applications, including motorcycle frames, roll cages and the inner structure of airplanes.

Pipe

Pipe is generally much heavier than tubing. Pipe is structural steel that usually carries gases or liquids. Most pipe is low-alloy or low carbon steel. Pipe can also be stainless steel, copper, aluminum, nickel steel, brass or chromium-molybdenum. Pipe is manufactured in nominal sizes ranging from 1/8 in. to 12.0 in. The Nominal Pipe Size (NPS) is a standardization of pipe measuring. The inside diameter of the pipe is determined by the wall thickness of the pipe and is referred to as the schedule of the pipe.

One thing we need to understand about the schedule of the pipe is that as the diameter increases, the schedule increases. Pipe is identified by schedule, that is, schedule 40, 60, 80, 120 and the like. The higher the schedule, the thicker the pipe and the more pressure it can withstand Pipe also comes in standard (STD) thickness, extra strong (XS), and double extra strong (XXS). The difference between these materials is the inside diameter or wall thickness. XS and XXS have much thicker walls, which gives them extra strength.

Fig 9.8 Pipe welding can be automated (top) or done manually (bottom) as seen here.

Bar Stock

Bar stock comes in shapes that include round bar, flat bar, hexagon bar and rectangular bar. Bar stock measurements are given in length x width. Bar stock is used for reinforcement and for machining purposes.

Symbol	Square Size		Length
⌀	2"	x	10'5"

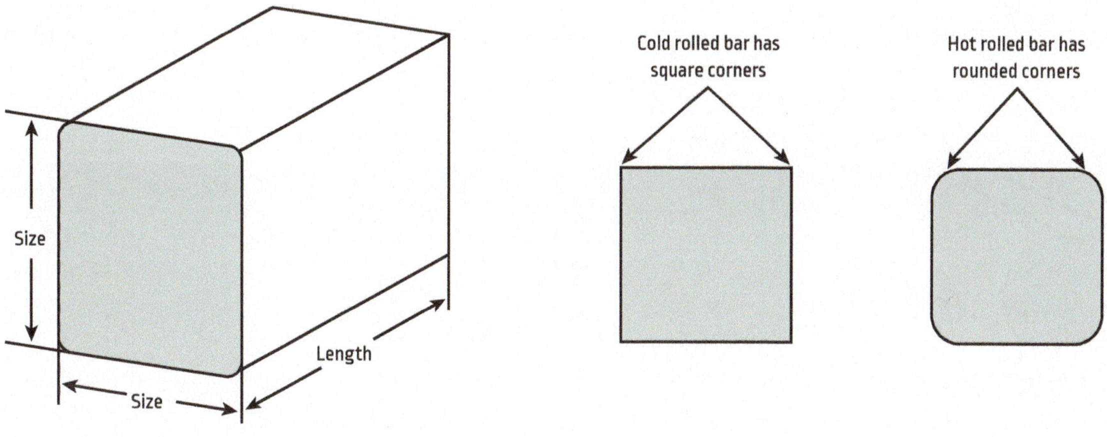

Figure 9.9 Round, flat and rectangular bar stock and measurements.

Flat Bar	Symbol	Width in Inches		Thickness in Inches		Length in Feet
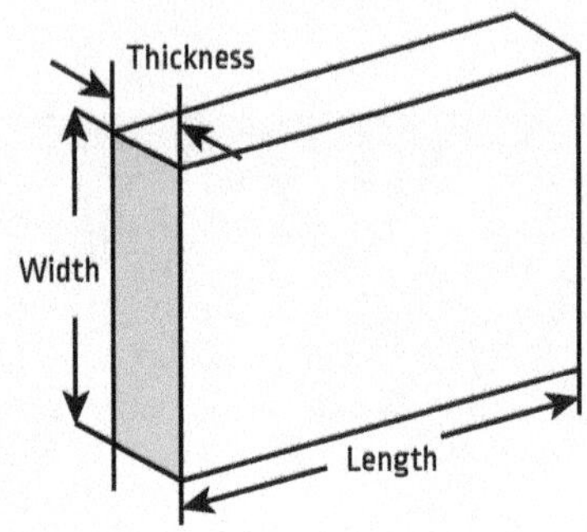	Bar	6"	x	3/8"	x	10'

Hexagonal Bar	Symbol	Distance across Flat Sides		Length in Feet
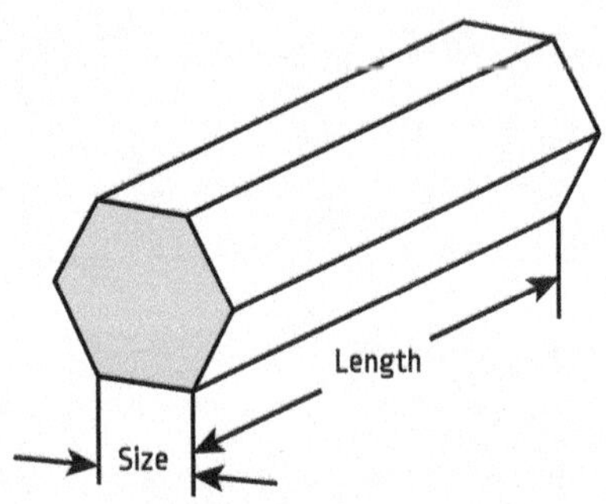	Hex	1"	x	12'

Octagon Bar	Symbol	Distance across Flat Sides		Length in Feet
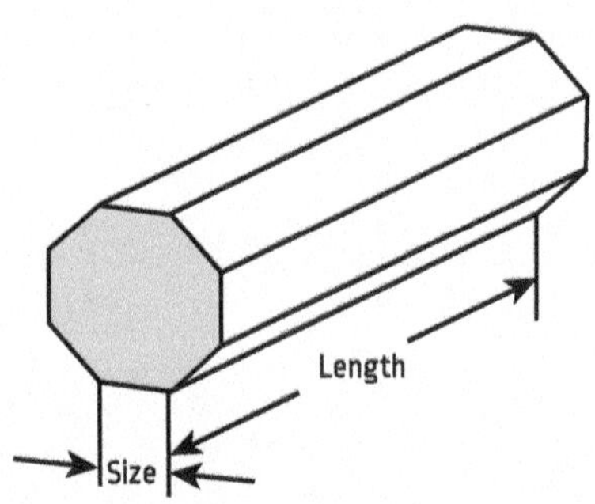	Oct	2"	x	10'

Round Bar	Symbol	Diameter		Length in Feet
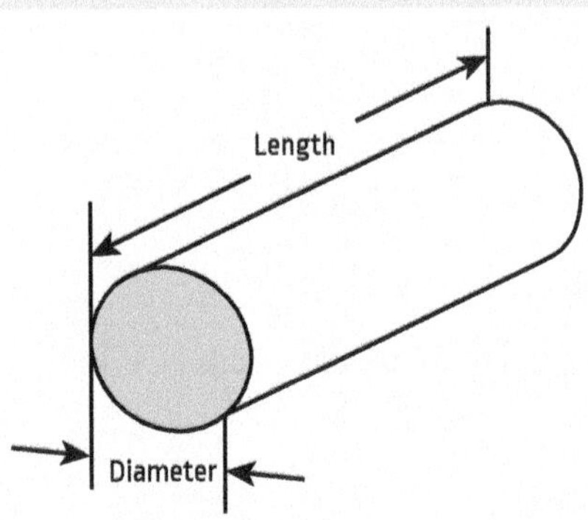	Ø	2"	x	11'

Figure 9.9 (Continued). Round, flat and rectangular bar stock and measurements.

Plate

According to the AWS, steel 3/16 in. and thicker is considered plate steel. Plate steel is used in many different capacities, including reinforcement members, parts and others. Plate steel is measured in length x width x thickness. Plate steel is defined by a specific manufacturing process in addition to its thickness.

Plate Steel	
Structural Shape	Symbol

Note: Materials thicker than 3/16" are referred to as "plate"

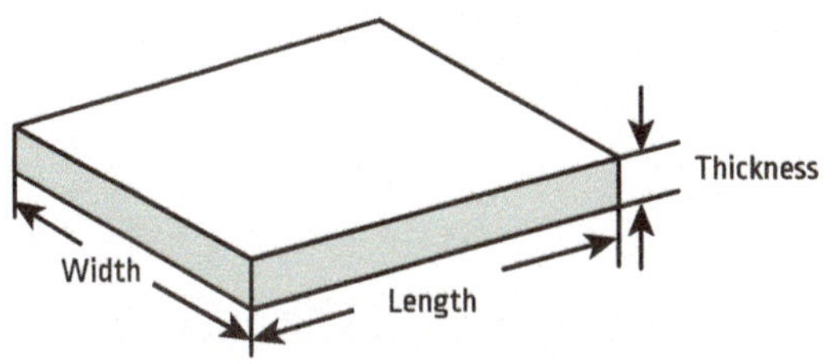

⅊

Size Specification				
Width in Inches		Thickness in Inches		Length in Feet
10"	x	1/4"	x	12'

Figure 9.10 Plate steel and how it is measured.

Sheet Steel

Steel 3/16 in. or thinner is considered sheet steel, and there are several types of manufactured sheet steel. Sheet steel is used extensively in the automotive and aerospace industries. It is used along with plate steel in the construction and shipbuilding industries and to construct certain vessels. It is used in conjunction with structural steel to add strength. Sheet steel can be reinforced with certain forming and shaping techniques. Its measurements are given in length x width x thickness in decimal or gage format.

Sheet Steel	
Structural Shape	Symbol

Note: Materials thinner than 3/16" are referred to as sheet steel

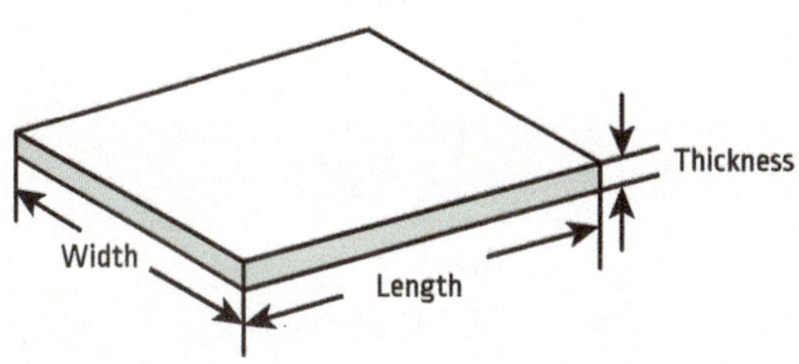

Sheet

Size Specification				
Width in Inches		Thickness in Inches		Length in Feet
24"	x	1/8"	x	8'

Figure 9.11 Sheet steel and how it is measured.

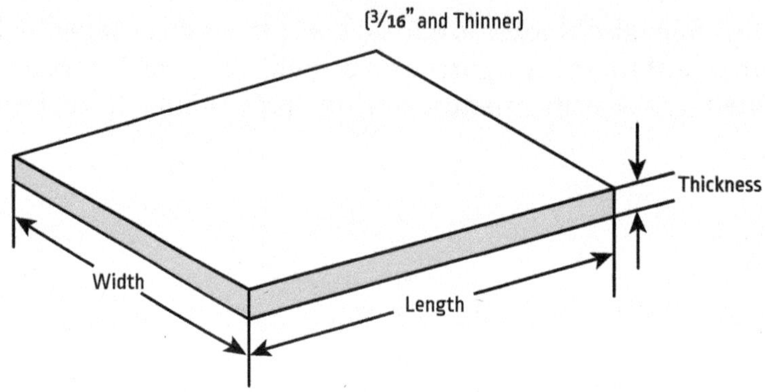

Note: Strip or band is measured the same as sheet steel. Strip or band is less than 3/16" thick and does not exceed 6" in width.

Figure 9.11 (Continued) Sheet steel and how it is measured.

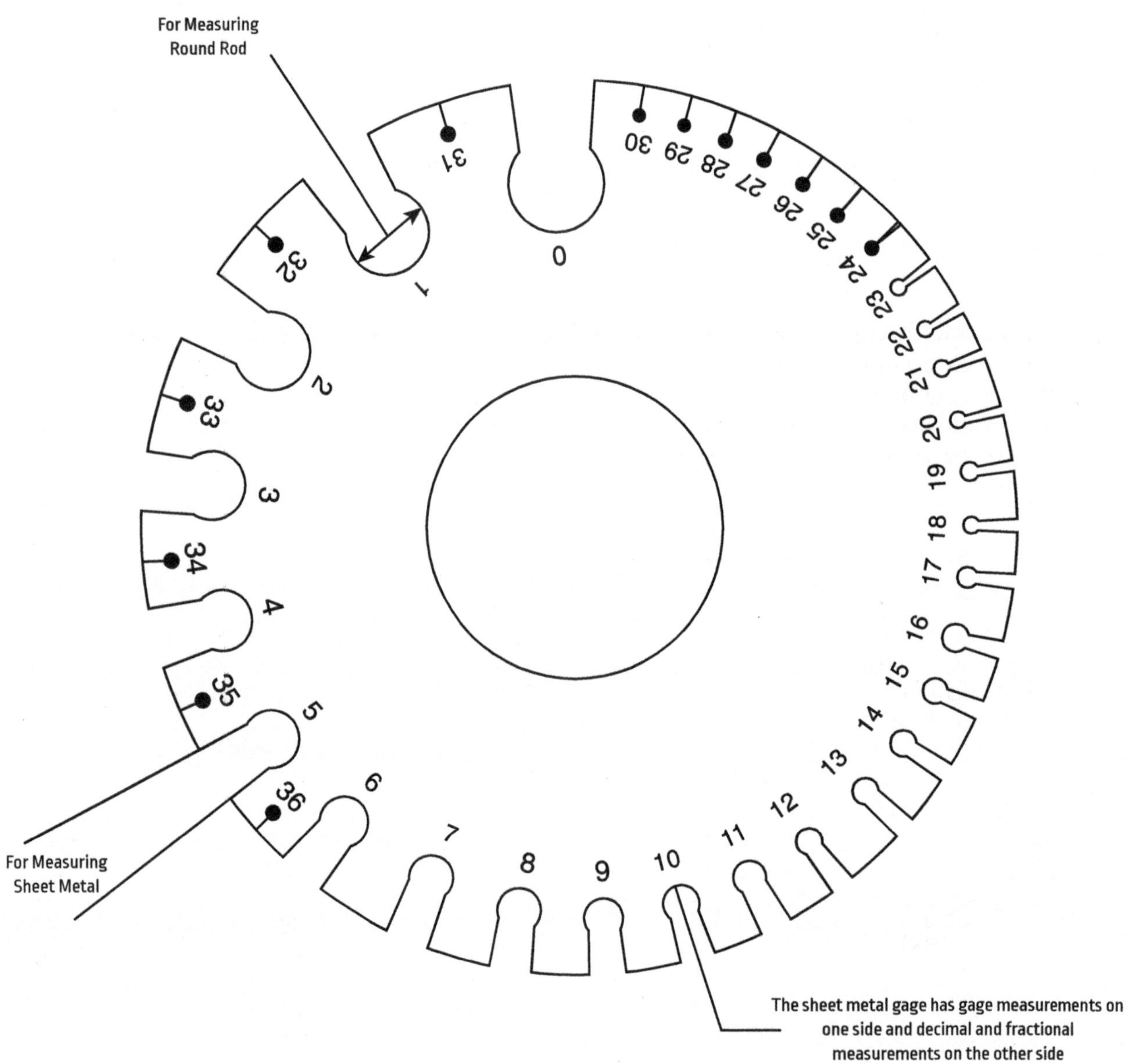

Figure 9.11 (Continued) Sheet metal gage.

Review Questions

Identify the following with proper name.

1.

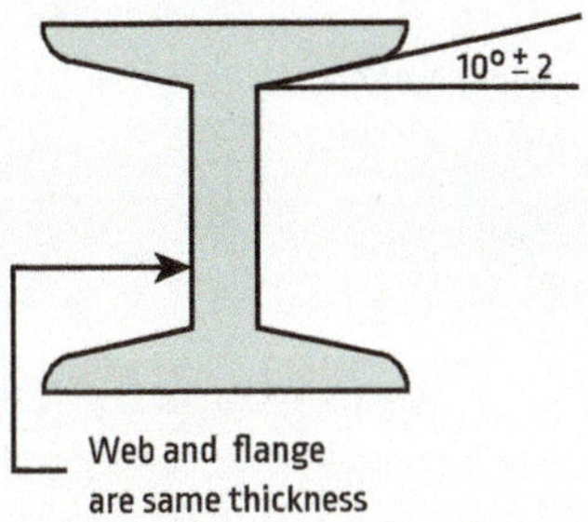

Name:

Symbol Designation:

2.

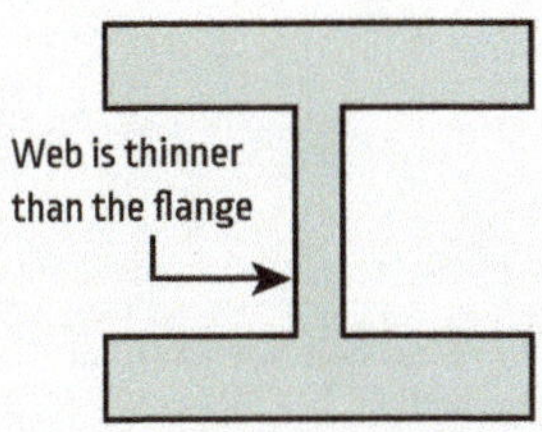

Name:

Symbol Designation:

3.

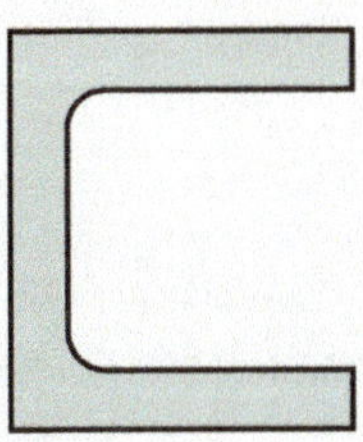

Name:

Symbol Designation:

4.

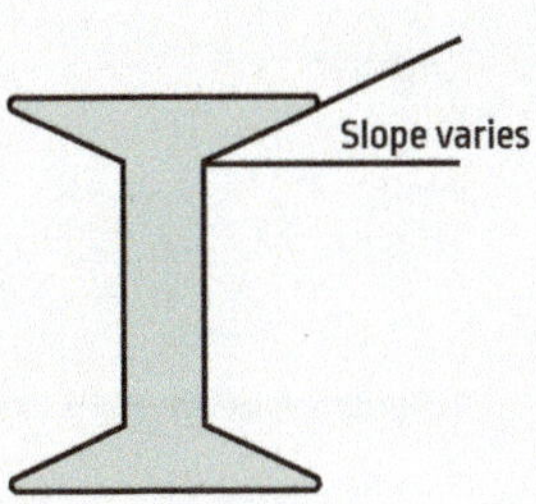

Name:

Symbol Designation:

5.

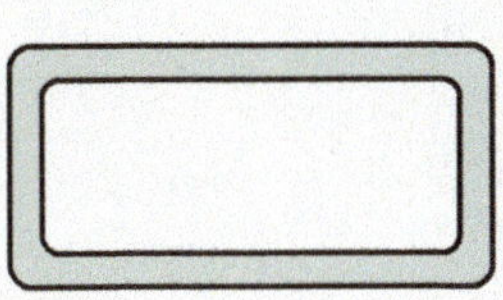

Name:

Symbol Designation:

6.

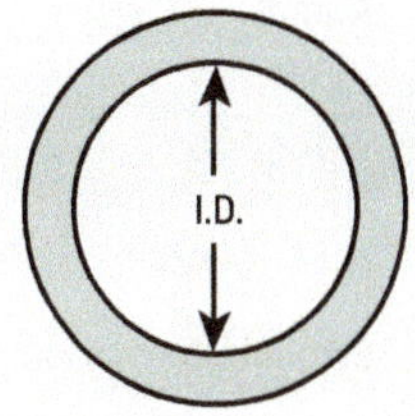

Name:

Symbol Designation:

7.

Name:

Symbol Designation:

8.

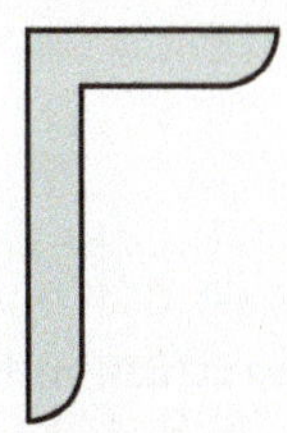

Name:

Symbol Designation:

9.

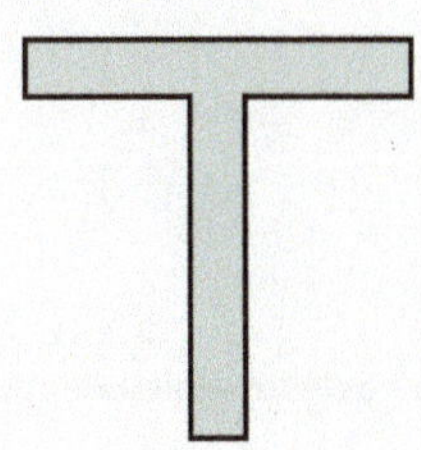

Name:

Symbol Designation:

10.

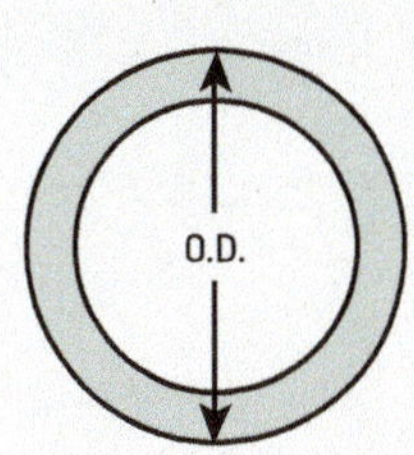

Name:

Symbol Designation:

Review Questions

Identify the following with proper symbol designation and name.

11.

Name:

Symbol Designation:

12.

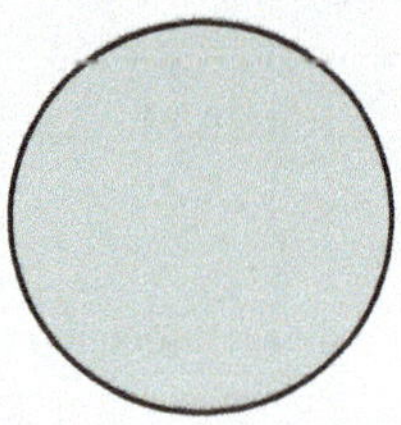

Name:

Symbol Designation:

13.

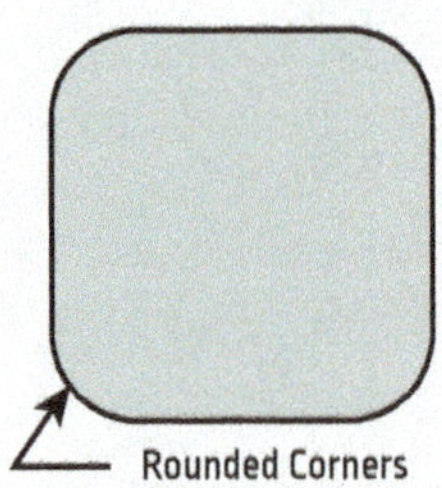

Name:

Symbol Designation:

14.

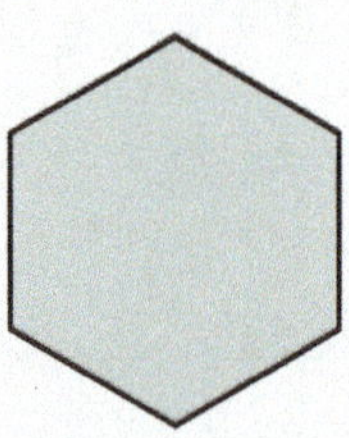

Name:

Symbol Designation:

15.

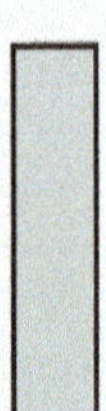

Name:

Symbol Designation:

16.

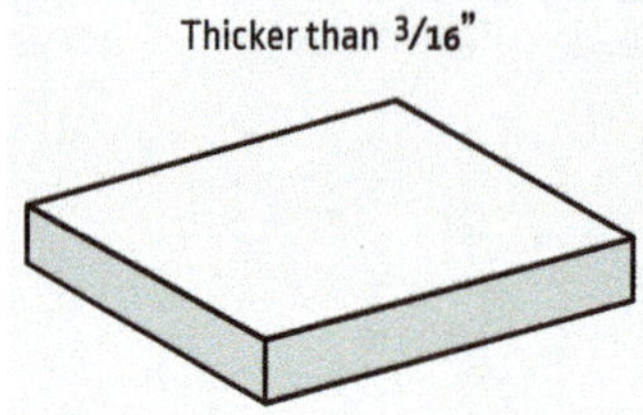

Name:

Symbol Designation:

17.

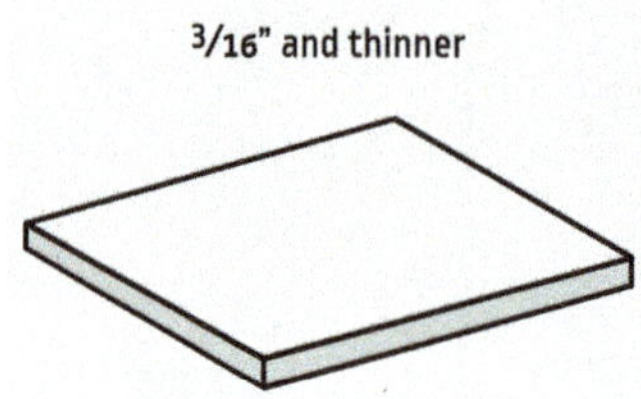

Name:

Symbol Designation:

18.

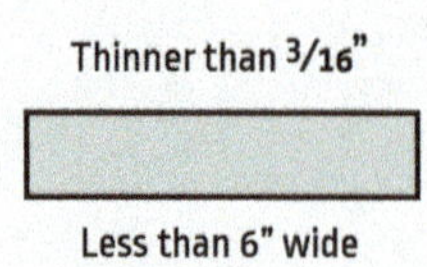

Name:

Symbol Designation:

19.

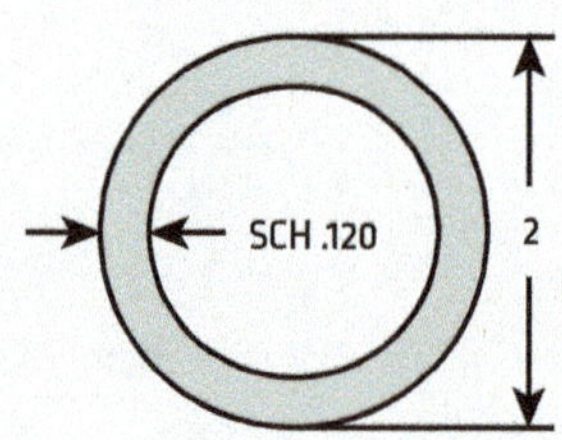

Name:

Symbol Designation:

20.

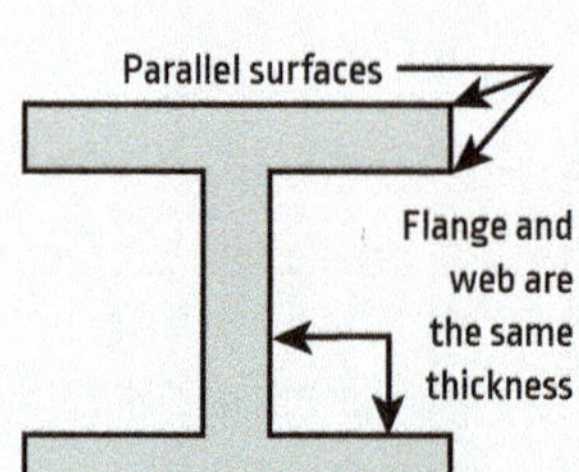

Name:

Symbol Designation:

Notes

Notes

CHAPTER 10: Weld Inspection & Examination

Chapter Objective

After learning the information in this chapter you will be able to:

- Understand the role welding codes play
- Identify destructive testing symbols and techniques
- Identify non-destructive testing symbols and techniques
- Describe the most common types of inspection
- Understand the inspection technique abbreviations

Key Terms

Non-Destructive testing
Destructive testing
Inspection symbols
Visual inspection (VT)
Penetrant testing (PT)
Magnetic particle (MT)
Ultrasonic (UT)
Radiograph (RT)
Eddy current (ET)
Leak test (LT)
Proof test

Quality control is a necessity in the competitive manufacturing market. There are many components to quality control, each of which can increase production, increase consistency or reduce the cost. Weld inspection and testing are two of the most effective ways to control quality in any welding or fabrication facility.

Quality levels vary with every weld shop and every product, and manufacturers who can maintain quality while producing a high volume of products at a competitive cost have an advantage over the competition. Many products must be built to strict requirements, or codes. If you do not understand welding inspection and testing methods, it will be difficult for you to adhere to a specific code. Our discussion of welding inspection and examination will be based on welding codes and examination criteria including the AWS, the Canadian Welding Bureau (CWB) and the American Society of Mechanical Engineers (ASME). The AWS and the CWB have developed welding codes for almost any conceivable situation.

Welding Codes

A welding code is a document used to ensure high quality by providing detailed directions for making a weld or building the part. Though there are many codes, a few common welding code bodies are the AWS D1 Structural Welding Code, ASME, and API. When a question regarding code arises, it is usually addressed by the AWS. In other cases, code interpretation is usually done by a certified welding inspector (CWI) who has learned and studied the welding codes and taken a certification test to prove competency with the code. Inspection criteria are clearly laid out in each code.

Destructive Examination

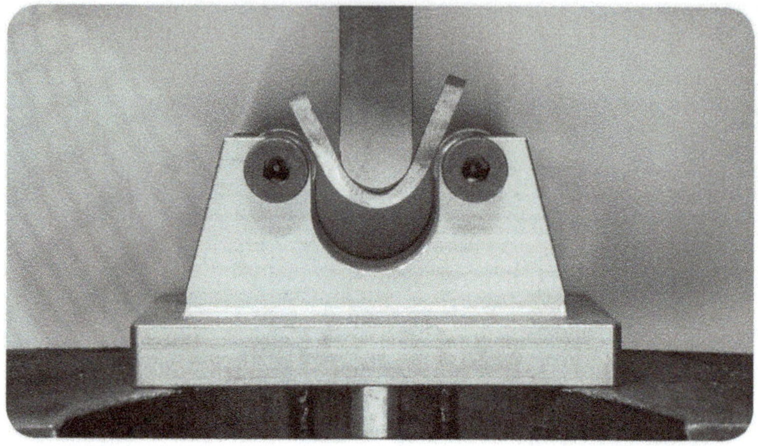

Figure 10.1 A guided bend test machine.

The term "destructive testing" is self-explanatory. In destructive testing, we actually destroy the weld to test it. Destructive testing includes guided bend tests, tensile pulls and fillet weld break tests. Destructive testing is usually done in a facility as part of the welder's performance qualification test or as part of a procedure qualification test. Welder qualification tests are done in accordance with the welding code the shop uses. The code provides acceptance criteria that the weld must meet in order to pass the qualification test.

Figure 10.2 Left, Bend test failure. Right, Bend test pass.

Non-Destructive Testing: NDT

When we're working in the field, we don't usually have the equipment to perform destructive testing. This is where non-destructive testing (NDT) takes over. For clarity, NDT, non-destructive examination (NDE) and non–destructive inspection (NDI) are interchangeable expressions or phrases that refer to the same process. There are several types of NDT methods, each with its own strengths and weaknesses. The code will determine which NDT method must be used. NDT can be used on one side or both sides of a joint. Performing NDT on several points of a weld can reduce the potential for failure, ensure quality and limit rework. This chapter will explain the basics of the NDT methods to help you understand weld examinations.

Welding Symbols

The correct NDT method will be indicated by a symbol on the weld drawing. The NDT weld symbol can be placed on either side of the reference line. Sometimes it's necessary to perform different types of testing on the same weld.

Symbol	Symbol Description
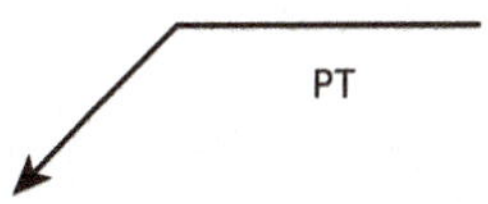 PT	Penetrant Test Arrow Side
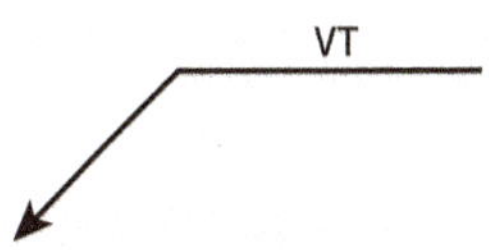 VT	Visual Test Other Side
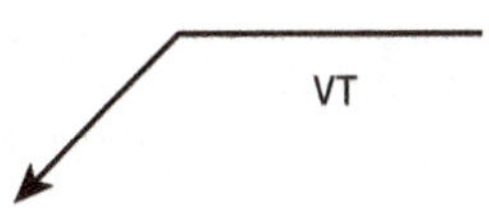 VT	Visual Test Arrow Side
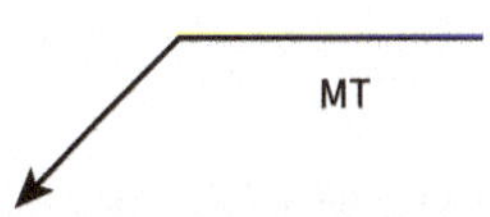 MT	Magnetic Particle Test Arrow Side
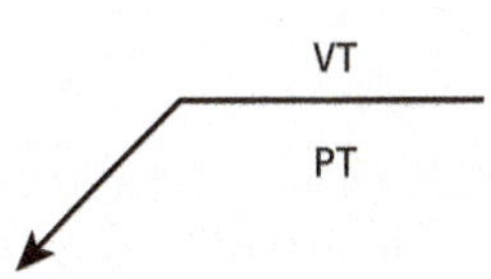 VT PT	Visual Test Other Side Penetrant Test Arrow Side
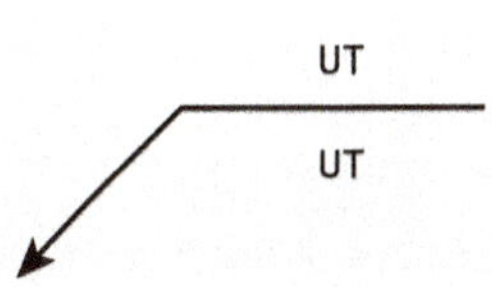 UT UT	Ultra Sonic Test Both Sides

Figure 10.3 NDT symbols.

Symbol	Symbol Description
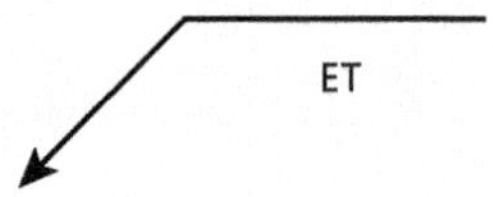	Electromagnetic Testing Arrow Side
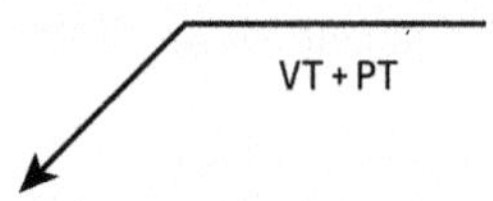	Visual Test Combined with a Penetrant Test Arrow Side
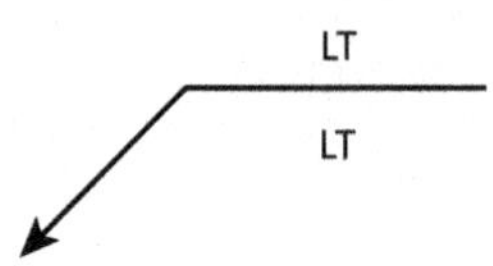	Leak Test Both Sides
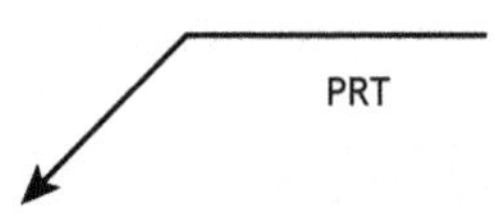	Proof Test Arrow Side
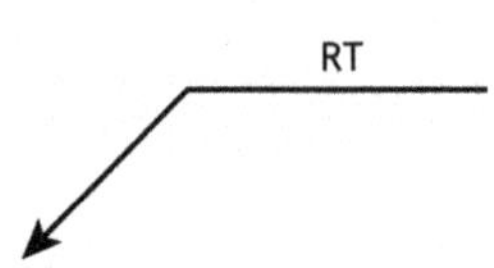	Radiographic Test Other Side
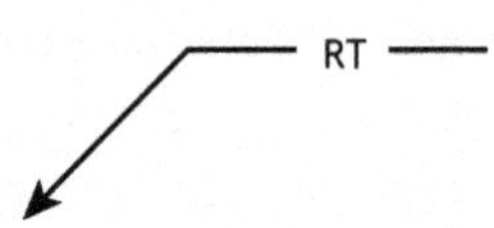	Radiographic Test No Side Significance
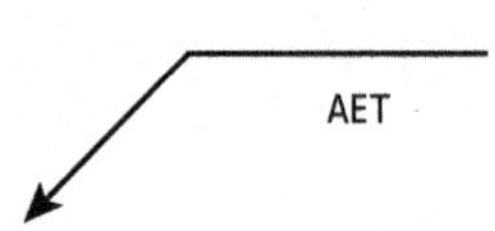	Acoustical Test Arrow Side

Figure 10.3 (Continued) NDT symbols.

Visual Testing (VT)

Every weld examination starts with visual testing. After a weld is completed in the field, a CWI performs a visual test to determine whether the weld fits within the product guidelines. Visual testing is the most common, most effective and least costly of all inspection processes. Visual testing can be done before welding, during welding and after the weld is completed.

A skilled inspector can easily determine whether or not the weld meets the inspection criteria. Extensive training is required to develop this level of skill. A CWI performs visual testing on the entire weld to identify possible problems before moving on to the next step of testing.

Fig 10.4 Weld inspection.

Dye Penetrant Testing (PT)

Dye penetrant testing can be used as a finish testing method or to identify problem areas throughout the weld. To perform PT the part must be cleaned, and then a thin layer of red dye is sprayed onto the surface of a weld joint and allowed to soak into the weld. After the dye has had time to soak into even small imperfections, it is then wiped away with a cleaning agent that removes any excess.

After the excess dye is cleaned from the joint, the final step is to spray a developer over the inspection area. Developer is a chalk-like material that draws the dye out of the discontinuities. As the developer dries, any remaining red dye is revealed, showing the imperfections in the weld. Anywhere there is a potential problem with the weld, a red dot will soak through the developer. Dye penetrant testing is a quick, accurate and economical testing method; however, it can only be used to detect surface discontinuities. We cannot use PT to see possible issues inside the weld as we can with some of the other methods.

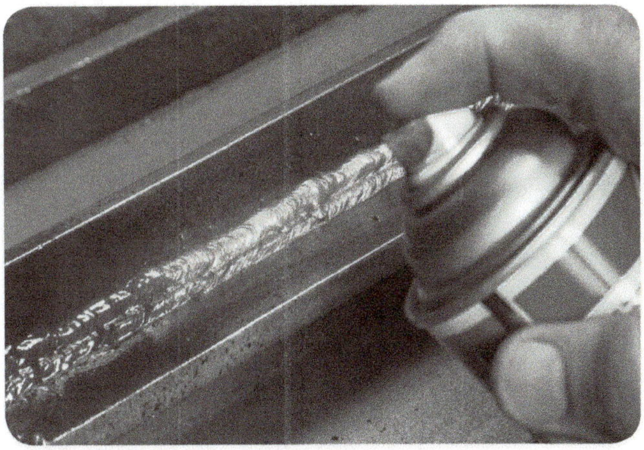

Figure 10.5 Dye penetrant.

Figure 10.6 Top, Apply developer to weld joint. Bottom, The developer reveals discontinuities.

Magnetic Particle Testing (MT)

Magnetic particle testing uses a magnetic field and iron powder to detect surface discontinuities and shallow surface discontinuities such as cracks in a part. It is often used on engine cylinder heads to detect cracks prior to the rebuilding process. Magnetic particle testing uses a device called a yoke to produce a magnetic field within the part. The yoke can create the magnetic field using AC or DC current. Colored iron powder is then applied to the surface and gathers near surface flaws and shallow subsurface flaws. Different colored powders can be used to provide contrast when inspecting a part. Using powder to perform MT is called the dry method. It is quick, accurate and easy to clean up afterwards. Magnetic particle testing can also be performed using a liquid in which the magnetic iron powders are suspended. This is known as the wet method. Fluorescent material can be added to the particles, which enables us to inspect the part using an ultraviolet light to reveal discontinuities.

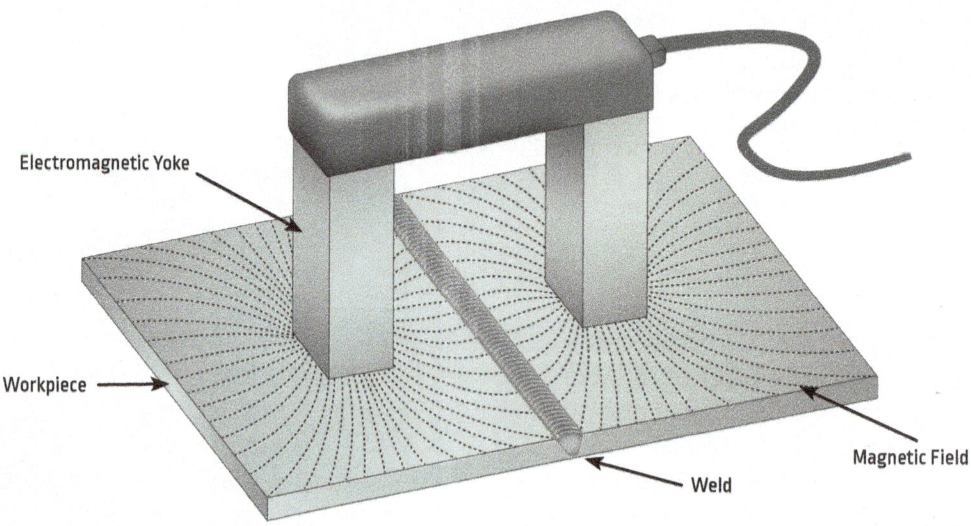

Figure 10.7 Magnetic particle testing.

Ultrasonic Testing (UT)

Ultrasonic testing sends sound waves into the material to detect internal discontinuities in the base metal or weld. Sound waves are sent through the part with a transducer and then reflected back to another receiver (although many transducers both send and receive the signal). The sound waves are displayed on the screen of the UT machine. If there are no issues with the weld, the sound waves will have a consistent shape. If there are any discontinuities in the weld, there will be an interruption in the shape of the waves. This examination process shows issues all the way through the weld, not just surface discontinuities. Ultrasonic testing is unique because it can indicate not only where a discontinuity is in the weld but also how deep within the weld the discontinuity is located. This is critical in determining exactly where a discontinuity is in a weld.

Structural Shape	Size Specification
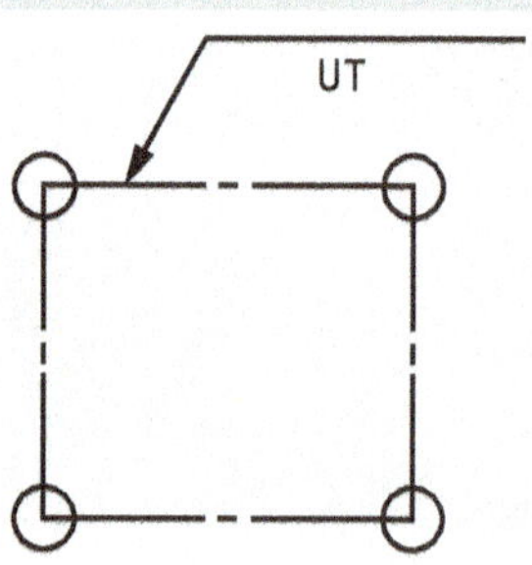	Ultrasonic inspection on the arrow side of the weld joint. Inspect the area enclosed by the boundary.

Figure 10.8 Symbol for ultrasonic testing.

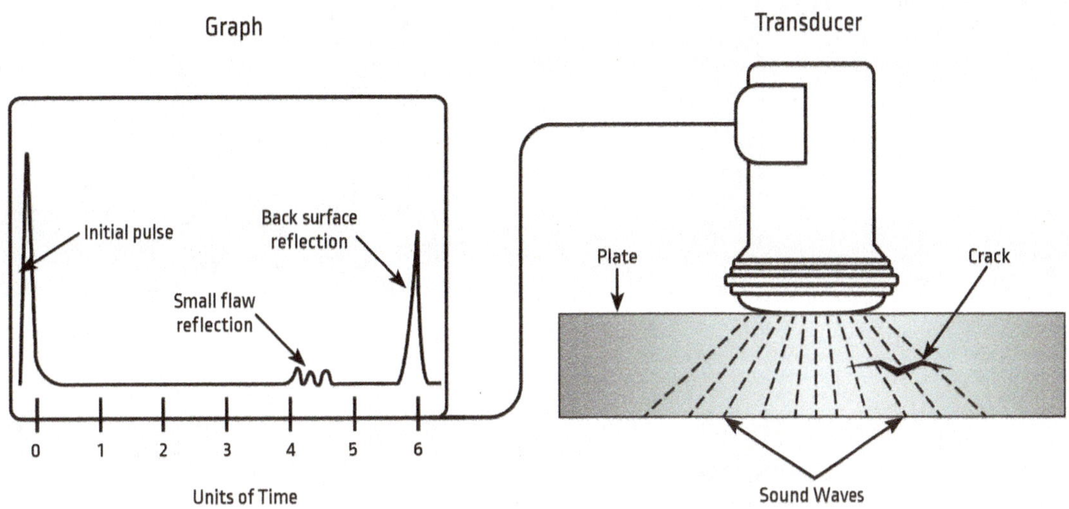

Figure 10.9 Ultrasonic testing setup.

Radiograph Testing (RT)

Radiographic testing is commonly known as X-ray testing. RT is done the same way as an X-ray of a broken bone. Radiographic testing is a dependable and accurate way to find internal discontinuities in a weld. It can be performed in the plant or in the field with portable equipment. The X-rays produce an image on the film, which must be interpreted by a certified RT inspector. The film will show the weld, and if there is any discontinuity, it will show up on the film as a light spot or a dark spot, line or area depending on the type of discontinuity. One issue with using this method is that radioactive materials are needed to perform RT, so exposure to radiation is a concern for the technician performing the testing. Because the testing materials are expensive, technicians must handle them carefully. They must also be sure to dispose of the materials correctly.

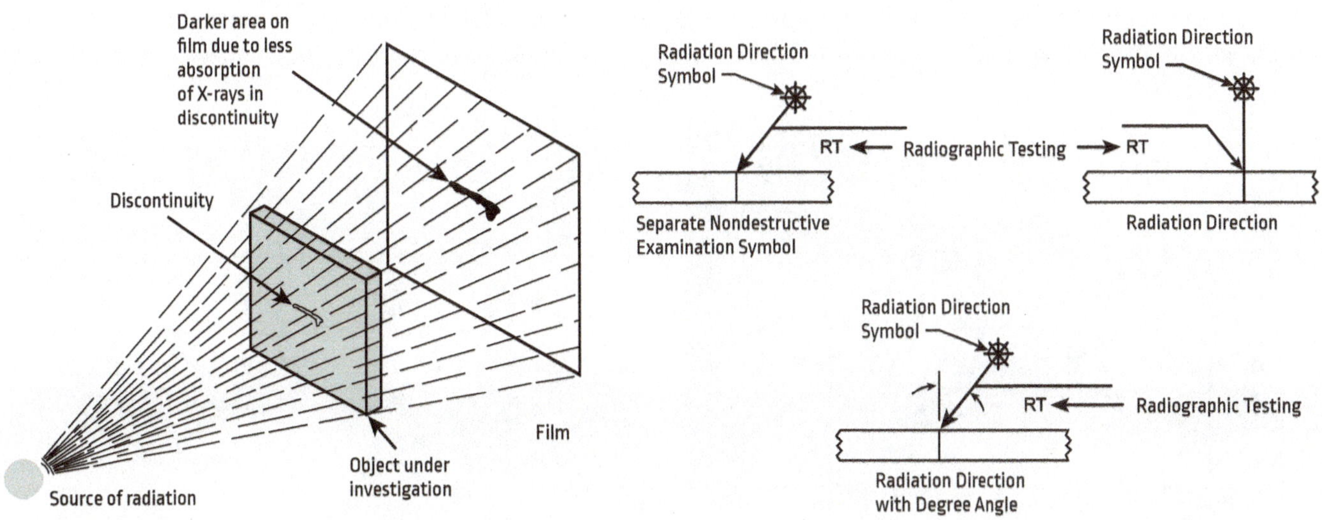

Fig 10.10 Radiograph testing (X-Ray) is a very accurate and dependable testing method used in many different aspects of this industry.

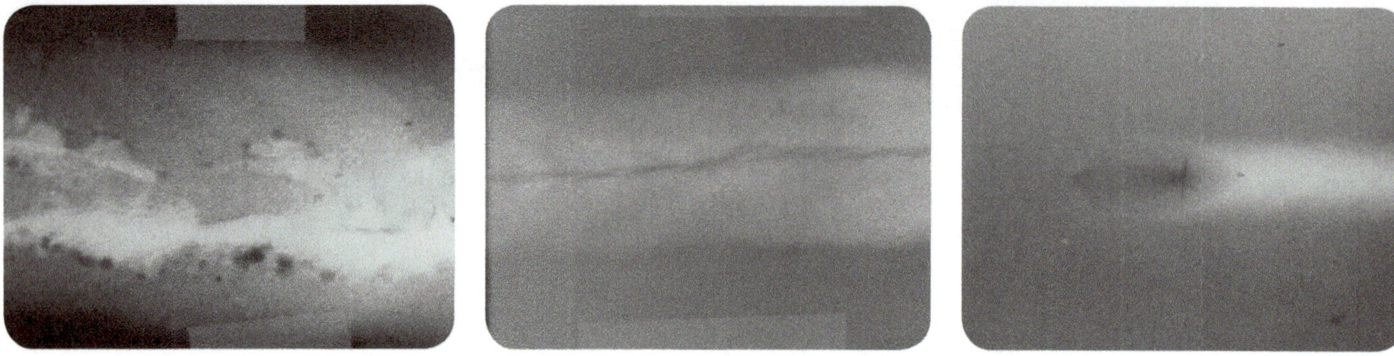

Figure 10.11 Radiographs. Left, Copper inclusion. Middle, Crack. Right, Crater crack.

Eddy Current Testing (ET)

Eddy currents work best on smooth surfaces and will show surface discontinuities just under the surface, such as cracks or porosity. Because eddy current is produced by alternating electromagnetic fields, this method can only be used on materials that will conduct electricity. Eddy current testing is often used in high production situations. Eddy current can also be used to test through coated materials, as long as the base material is conductive.

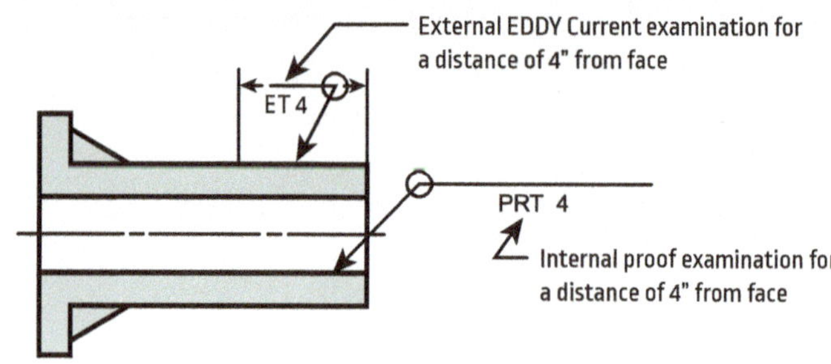

Figure 10.12 Eddy Current symbol

Leak Testing (LT)

Leak testing, commonly known as hydrostatic testing, is usually performed on tanks or pipes. To perform LT, the vessel being tested is filled with water, and pressure is applied. Depending on the application, the pressure can range from 5.0 pounds per square inch (psi) to several thousand psi. After pressure is applied, the vessel is visually inspected for leaks. Any leaks are identified, fixed and re-tested. It is dangerous and therefore not recommended to test a tank without filling it with water first.

Another way to test low pressure tanks is to pressurize the tank and then apply soapy water to the outside of the tank. Soapy bubbles will form where there are holes in the tank. Leak tests are both simple and effective.

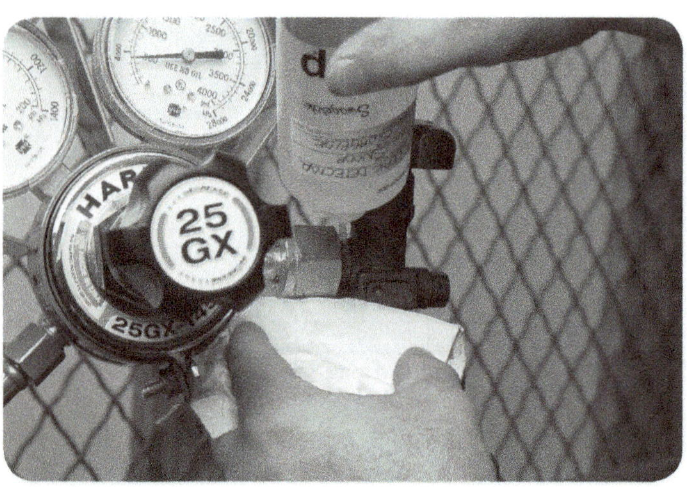

Figure 10.13 Apply soapy water or leak detection fluid to check for leaks.

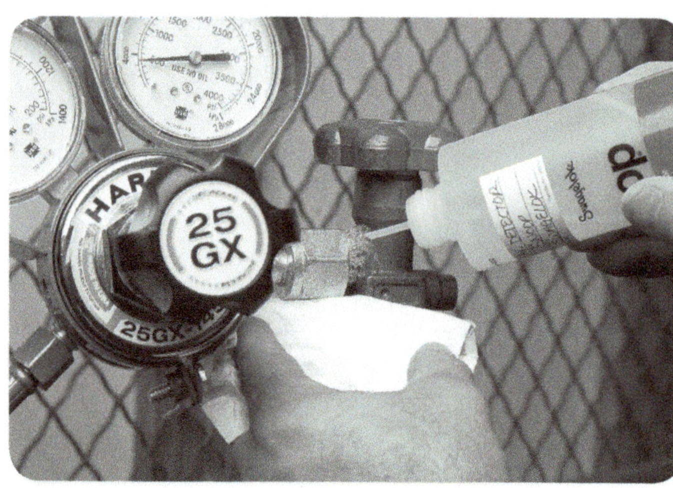

Figure 10.14 The soapy bubbles indicate a leak.

Proof Testing (PRT)

Proof testing provides proof that the part or vessel has been constructed to a specific standard. In proof testing, the pipe system, vessel or tank is pressurized well above the normal working conditions to which the part will be subjected. For example, proof testing is performed as part of the manufacturing process for natural gas compressors. The pipes in these compressors will typically be subjected to pressures of 5,000 to 6,000 psi.. The company requires that testing be done at 8000.0 psi. This ensures that the welds can easily handle the pressure ranges to which they'll be subjected every day.

The type of NDT used depends on several variables. Sometimes the welding code will specify the type of testing to be done, sometimes the engineers will determine the level of necessary testing and sometimes testing is governed by budget. Non-destructive testing requires inspectors who are well-versed in the testing process and are able to determine if the weld meets the inspection criteria or if it needs to be fixed. The AWS offers training and certification testing in several of the more complicated NDT processes—another avenue your welding career can take you in if you are willing to put in the work.

Review Questions

Describe the weld symbol.

1. 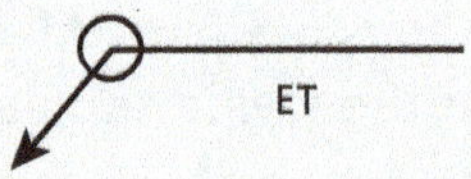

2. 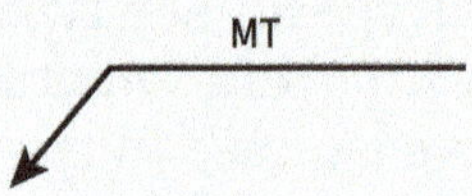

3. 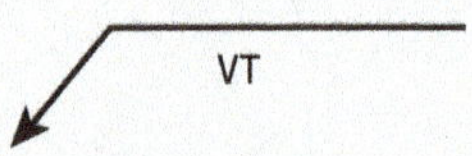

4. 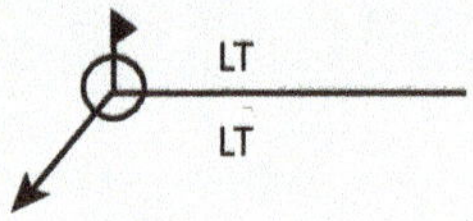

5. 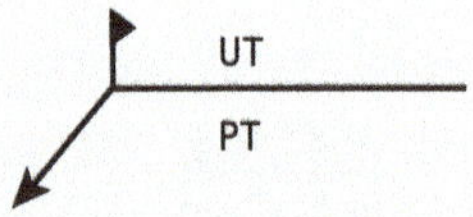

6. 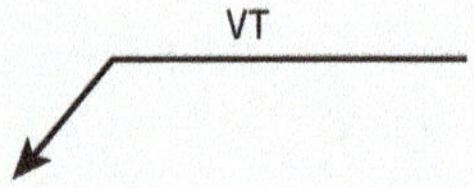

7. 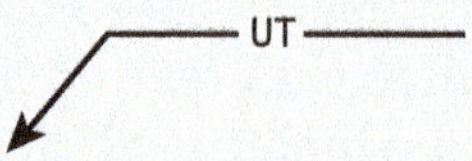

8. 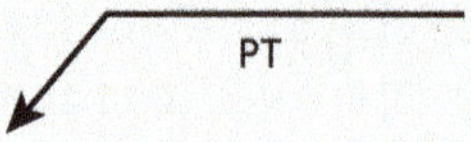

9. 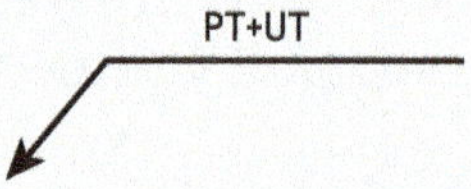

10. 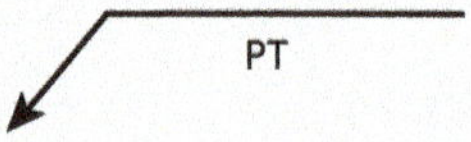

11. 

12. 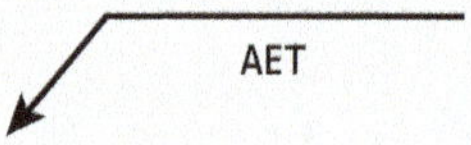

True/ False Questions. Check the correct answer.

1.	A defect is a term used to describe a rejectable weld, due to not meeting inspection criteria.	True	False
2.	A discontinuity is a flaw in the weld but still meets the inspection criteria.	True	False
3.	Visual testing is only used in addition to destructive testing.	True	False
4.	Multiple nondestructive examinations can be required for the same weld.	True	False
5.	Destructive examinations usually destroy the weld physically.	True	False
6.	Ultrasonic testing uses X-rays to inspect the inside of the weld.	True	False
7.	Dye penetrant testing is used to detect surface impurities.	True	False
8.	A radiograph uses radiowaves to detect flaws.	True	False
9.	X-ray inspection uses film and a radiation source.	True	False
10.	A leak test usually requires a minimum of 100 PSI.	True	False

Notes

Notes

Notes

Notes

Notes

Notes

Notes

Notes

Glossary

acoustic emission testing- A non-destructive testing method that measures the amount of change in a part after it is subjected to a difference in atmosphere (hot and cold), when stress is applied, or increased or decreased pressure in a tank or vessel.

alloy- A substance with metallic properties and composed of two or more chemical elements of which at least one is metal. Chrome-molly and stainless steel are examples.

angle- A figure formed by two lines meeting at the same point. Angles are typically measured in degrees.

arrow side- The side of the weld joint to which the arrow points.

auxiliary view- Additional views in a drawing showing a surface (typically slanted) that cannot be seen from the top, front or right side view.

back gouging- The process of removing material from the back side of a weld before welding from that side. This ensures full fusion of the weld and the back weld into one weld.

back weld- A weld made at the back of a single groove weld, used to ensure full penetration with less joint preparation. Back welds are completed after the gouging process is complete and after the weld is performed.

backing bar- A piece of metal used on the backside of a weld joint to eliminate open root joints. Backing bars keep the weld from falling through the backside of the joint. Typically used in welder qualification testing.

backing weld- Weld that is performed on the backside of the joint before the actual weld is made. Backing welds serve the same purpose as backing bars.

bar stock- Solid raw material that can come in many different configurations, such as round bar, flat bar, hexagon bar, octagon bar, and hot rolled or cold rolled square bar.

base metal- The material being welded, brazed, soldered or cut; also referred to as the "parent" metal.

beam- A long piece of metal typically used in the structural steel industry. There are many different types, including I-beams, standard beams, wide flange beams and bearing pile beams.

bevel angle- The angle applied to a base metal by some means of cutting or machining to allow access to the root of the weld joint, measured in degrees.

bevel-groove- A weld joint in which one of the members has a single or double bevel. The print specifies which member, bevel angle, root opening and groove depth.

both sides- An indication in a weld symbol that a weld should be completed on both sides of the weld joint.

brazing (B)- The process of fusing metal with an oxyfuel torch. Brazing operates above 840° F but does not melt the base metal.

C- The abbreviation for a commonly used structural steel shape produced in the shape of a C with squared edges; also referred to as C-channel.

center line- The indication of the center of a hole, circle or arc.

chain intermittent fillet welds- Fillet welds made on both sides of a joint at the exact same location on each side.

chamfering- The process of angling the sides of a hole to create a cone shape, which gives access to the bottom of a hole. This is also done to create angles on parts such as round stock.

combination weld- Multistep weld that requires more than one joint type or process.

complete joint penetration (CJP)- The condition in which weld metal protrudes all the way through the material, ensuring 100% fusion; also referred to as full penetration.

consumable inserts- Filler metal placed at the joint root before welding and intended to be completely fused into the joint root to become part of the weld.

converting measurements- The process of switching measurements to a different measuring system; e.g., millimeters to inches, inches to decimals, inches to millimeters.

counter sink- A cone-shaped recessed hole that allows a fastener to sit flush with the base material.

crosshatching- A series of intersecting lines that show a cutaway of a part or distinguish different types of materials on a print.

cutting plane line- Line that shows internal features of a part that cannot be seen from a given view.

decimal- A number system based on the number ten or tenths. Decimals are used when measuring in the metric system.

destructive examination- A weld test in which the specimen is cut apart and physically destroyed as part of the testing process.

dial calipers- A device used to accurately measure outside diameter, material thickness, depth of holes, inside diameters, spacing and steps in the material.

dimension line- A line used to define the measurement of a part feature. Consist of a solid line with arrows at both ends and a dimension at the center.

dimensioning- The process of adding specific measurements to each component of a part.

dual dimensioning- Dimensioning in which measurements are expressed in both fractional and metric measurements on a drawing.

dye penetrant testing- A non-destructive testing method that uses a dye and a developer to identify any surface discontinuities.

eddy current testing (ET)- A nondestructive testing method that identifies surface discontinuities such as porosity using alternating electromagnetic fields.

edge weld- A weld in an edge joint, a flanged butt joint or a flanged corner joint in which the full thickness of the members are fused.

extension line- A line used to visually connect the ends of a dimension line to the relevant feature on the part. Extension lines are solid and are drawn perpendicular to the dimension line.

fillet welds- A weld of approximately triangular cross section joining two surfaces approximately at right angles to each other in a lap joint. Fillet welds can be single pass or multipass depending on the material thickness.

flare bevel-groove weld- A groove weld consisting of a member with a curved surface and another with a planar surface.

flare V-groove weld- A groove weld consisting of two members with curved surfaces.

flux cored arc welding (FCAW)- An arc welding process that uses a continuously fed consumable electrode that contains flux at its core. There are two types of Flux Cored wire: gas shielded, which uses additional gas for weld coverage, and self-shielded, which does not use a cover gas.

fraction- A unit of measurement. Typically used in the U.S.A representation of a part of a whole. Fractions are used when measuring in the empirical system.

gas metal arc welding (GMAW)- An arc welding process that produces coalescence of metals by heating them with an arc between a continuous filler metal electrode and the workpieces. Shielding is obtained entirely from an externally supplied gas.

gas tungsten arc welding (GTAW)- An arc welding process that produces coalescence of metals by heating them with an arc between a tungsten electrode (non-consumable) and the workpieces. Shielding is obtained from a gas. Pressure may or may not be used, and filler metal may or may not be used.

hardfacing- A surfacing variation in which surfacing metal is deposited to reduce wear.

hidden lines- A line used to define a part feature not visible in a specific view. Hidden lines consist of a series of short dashes.

included angle- The angle between two sides, formed when two lines meet at a common point.

intermittent fillet weld- A fillet weld in which continuity is interrupted by recurring unwelded spaces.

intermittent groove weld- A groove weld in which continuity is interrupted by recurring welded spaces. These welds are based on weld length and center-to-center measurements called "pitch."

ISO- The International Organization of Standardization.

isometric view- A three-dimensional view created from the top, front and right side view.

lap joint- A type of joint between two overlapping metal parts in parallel planes. Lap joints are welded with fillet welds and can be single pass or multipass welds depending on the material thickness.

leader lines-end- A thin line with an open or closed arrow head that that is often positioned at an angle and used to tie a dimension to a feature, especially when there are space limitations.

leak testing (LT)- Typically used in vessels, tanks and piping, leak tests use water and pressure to identify any leaks. The pressure can range from 3-5 PSI to several thousand pounds of pressure.

long break lines- A line used to show long breaks in a part, usually done by hand.

magnetic particle testing- A non-destructive testing process that uses a magnetic field and iron powder sprinkled on the part to detect discontinuities. It is best used to detect surface cracks.

metric- An international system based on the meter as a unit of length.

multiple reference lines- A weld symbol that identifies the procedural steps needed to perform a weld.

multiview drawing- An illustration technique that uses top, front and right side views to describe a part on a print.

non-destructive testing (NDT)- Tests that assess the integrity of a weld without destroying the specimen (e.g., radiographic or ultrasonic testing).

non-destructive testing designation- The letter label for the type of non-destructive test to be performed.

object line- A solid line used to define the shape and size of a part feature.

orthographic projection- A drawing that communicates the shape and size of an object through a series of related two-dimensional views.

other side- The weld to be placed on the opposite side of where the arrow is pointing in a weld joint, completed according to the instructions given above the reference line.

oxyfuel welding (OFW)- A welding, heating and cutting process that uses a combination of a fuel gas and oxygen to create a flame hot enough to melt metal.

partial joint penetration (PJP)- A groove weld in which incomplete joint penetration exists.

phantom line- Dashed line that shows the movement of a part, such as a rotating plate, on a drawing.

pipe- Round tube made of various metals used to pass a material such as water, oil or gas through it. Pipe is measured by the inside diameter (ID).

pitch- The center-to-center distance between intermittent welds.

plate steel- Flat steel sheet over 3/16 in. in thickness.

plug weld- A type of weld made by joining one metal part with a circular hole to another metal part positioned directly beneath it.

process designation- The acronym given to each welding process (e.g., GMAW for Gas Metal Arc Welding or GTAW for Gas Tungsten Arc Welding).

projection welding- (PW)- A resistance welding process that produces coalescence by the heat obtained from the resistance to the flow of the welding current. The resulting welds are localized at predetermined points by projections, embossments, or intersections.

proof testing (PT)- A stress test that validates that the part has been constructed to a specific standard. This type of testing consists of putting the part into service conditions above expected working conditions.

protractor- A tool that measures angles.

radiographic testing (RT)- A non-destructive test that uses a radiation source and film to detect internal discontinuities in a weld. Also known as X-ray testing.

resistance seam welding (RSEW)- A resistance welding process that produces coalescence at the faying surfaces of overlapped parts progressively along a length of a joint. The weld may be made with overlapping weld nuggets, a continuous weld nugget or by forging the joint as it is heated to the welding temperature by resistance to the flow of the welding current.

resistance spot welding (RSW)- A resistance welding process that produces coalescence at the faying surfaces of a joint by the heat obtained from resistance to the flow of welding current through the workpieces from electrodes that serve to concentrate the welding current and pressure at the weld area.

resistance welding (RW)- A group of welding processes that joins parts by the heat obtained from the resistance to the flow of electric current. Pressure is applied to weld the parts together.

section line- Angled or crosshatched line that shows a difference in material (e.g., steel, aluminum, copper, cast iron).

sectional view- A view illustrating a rotated section resulting from an imagined cut in the part. Without sectional views, many hidden lines would be needed to describe the inside of a part, creating confusion.

sheet steel- Flat sheet steel 3/16 in. and thinner, typically measured in gauge thickness. If the thickness is over 3/16 in., it is referred to as plate steel.

shielded metal arc welding (SMAW)- An arc welding process that uses a flux-coated consumable rod electrode. It is also referred to as stick welding.

short break line- Line used to shorten dimensions that are excessively long, usually done by hand.

sketching- The process of creating a rough drawing to quickly convey thoughts in a visual form.

slot weld- A type of weld made by joining one metal part with an elongated hole to another metal part positioned directly beneath it.

soldering (S)- A joining process in which a filler metal is melted at temperatures below 840° F (450° C) to form a joint between two base metals. Soldering is often used for delicate projects such as jewelry and electronics.

spacer- A metal part, such as a strip, bar or ring, inserted into the joint root to serve as a backing and to maintain the root opening during welding. After tacking the parts the spacer is removed.

square groove- A type of weld that consists of an opening between two square part surfaces, which provides space to contain weld metal. Typically used for materials 3/16 in. and less.

staggered intermittent weld- An intermittent weld on both sides of a joint in which the weld increments on one side are alternated with respect to those on the other side.

strength requirements- Determined by the engineer or the contract provided to establish the needed strength of a weld to serve in a given application.

structural steel- Steel construction material generally thicker than 3/16 in. used to build buildings, bridges, railroads, heavy equipment and anything made of thicker sections of materials.

stud welding (SW)- A general term for joining a metal stud or similar part to a workpiece. Welding may be accomplished by arc, resistance, friction or other process with or without external gas shielding.

submerged arc welding (SAW)- An arc welding process that produces coalescence of metals by heating them with an arc or arcs between a bare metal electrode or electrodes and the workpieces. The arc and molten metal are shielded by a blanket of granular, fusible material on the workpieces. Pressure is not used, and filler metal is obtained from the electrode and sometimes form a supplemental source (welding rod, flux or metal granules).

surfacing weld- A weld applied to a surface, as opposed to making a joint, to obtain desired properties or dimensions. Worn parts are surface welded and re-machined to put the part back into service.

tubing- A round tubular member that transports a material; similar to pipe but measured by outside diameter (OD).

U- and J-groove- Groove weld typically used on thicker sections to allow access to the bottom of the joint. These weld joints are identified by the shape of the joint.

ultrasonic testing (UT)- A non-destructive testing method that uses sound waves to indicate any discontinuities in a weld. This testing is especially desirable because it can identify internal issues in the weld and does not require radioactive material as with an X-ray.

weld face- The exposed surface of a weld on the side from which welding occurs.

weld finishing- The process of completing weld to a given specification; may include chipping, grinding, machining, hammering, rolling or peening.

weld leg- The distance from the root to the toe of the fillet weld. The size of the fillet weld is determined by the length of its legs. A fillet weld has two legs, the vertical leg and horizontal leg.

weld root- The point at which the back of the weld intersects the base metal surfaces. In many welds, the root is the most important weld to make correctly.

weld size- The dimensions of a weld that include leg length, convexity, and concavity. Weld size is determined by measuring tools specifically designed for welds.

weld symbols- A graphic character connected to the reference line specifying the joint geometry or weld type. Weld symbols contain all the necessary information to perform a specific weld (e.g., weld placement, weld process and weld size and length) in a small area on a drawing.

weld throat- The distance from the weld face to the root of the weld.

weld toe- The point at which the weld face and the base metal meet.

Conclusion

Learning how to read and interpret blueprints and drawings is a critical skill for most professional welders. It is a growing skill that is necessary for companies to remain competitive in a global market. Any disconnect of information between engineers and the floor welder creates frustration, added work and, above all, added costs. Learning the language of print reading will make you a better welder and enable you to outperform welders who do not understand how to read a print.

Throughout the book we have given you the necessary tools and information to guide you into the world of print reading. As you employ these skills, reading prints will become easier and easier until you become proficient at reading and interpreting the information. The James F. Lincoln Foundation produced this book to help you develop your skill level in your chosen trade of welding. The foundation has been influential in helping the welding trade advance through educational publications since the 1930's, which is a long time to remain dedicated to a task.